高等学校教材

第四版

机械设计制造及其自动化专业英语

大学英语专业阅读教材编委会　组织编写

马玉录　刘东学　主编

安　琦　主审

化学工业出版社

·北京·

内 容 提 要

《机械设计制造及其自动化专业英语》共分为三部分（Part），30个单元（Unit）。第一部分为机械设计与制造的基本知识，包括金属材料、非金属材料、材料的机械性能、材料力学、金属材料热处理、机械及机械零件设计、制造工艺、加工设备、增材制造、特种加工工艺、数控机床、液压系统、热力学、流体力学、化工机器、质量保证与控制等。第二部分为自动控制的基本知识，包括控制原理、控制系统类型、反馈控制原理、测量系统、过程控制、传感器及信号转换、系统状态监测、人工智能、可编程逻辑控制器等。第三部分为现代先进制造技术，主要介绍机电一体化、计算机数控、机器人、计算机辅助制造、柔性制造系统、计算机集成制造、自动组装、敏捷制造、精益制造、大批量定制生产、虚拟制造、绿色产品制造等。

本书的内容覆盖了机械设计制造及其自动化专业的基本内容及最新发展成果。各单元之间，既有一定的内在联系，又独立成章，可根据不同学时数灵活选用。

本书可供机械工程及相关专业本科生使用，也可作为同等程度的专业技术人员的自学教材。

图书在版编目（CIP）数据

机械设计制造及其自动化专业英语/马玉录，刘东学主编．—4版．—北京：化学工业出版社，2020.7（2024.8重印）

ISBN 978-7-122-36792-1

Ⅰ．①机…　Ⅱ．①马…②刘…　Ⅲ．①机械制造-英语-高等学校-教材②自动化技术-英语-高等学校-教材　Ⅳ．①TH16②TP

中国版本图书馆CIP数据核字（2020）第080931号

责任编辑：高　钰　郝英华　　装帧设计：史利平

责任校对：王　静

出版发行：化学工业出版社（北京市东城区青年湖南街13号　邮政编码100011）

印　　刷：北京云浩印刷有限责任公司

装　　订：三河市振勇印装有限公司

787mm×1092mm　1/16　印张14½　字数355千字　2024年8月北京第4版第5次印刷

购书咨询：010-64518888　　售后服务：010-64518899

网　　址：http://www.cip.com.cn

凡购买本书，如有缺损质量问题，本社销售中心负责调换。

定　　价：46.00元

版权所有　违者必究

前言

《机械设计制造及其自动化专业英语》（第三版）于 2015 年 2 月出版以来，已先后重印了多次，受到了本专业和相关专业教师和学生以及专业技术人员的支持和爱护，目前已有近 20 所高校采用该教材。

根据我们的教学实践和听取其他高校在使用本教材过程中提出的宝贵意见和建议，在前一版的基础上对全书进行了修订和补充，进一步拓宽、更新、提升教材内容。第四版基本结构和格式上与第三版保持一致，根据教学需要更换和调整了部分内容，增加了如增材制造、特种加工工艺、人工智能、可编程逻辑控制器等内容，保留的内容尽可能使用最新版本的内容。教材内容既基本涵盖本专业本科生应掌握的专业知识，也反应机械制造行业近年来的发展成果。这样，使教材的内容具有更强的时效性和适用性。

本教材由马玉录、刘东学主编，修订工作主要由华东理工大学马玉录、周邵萍、洪瑛、李琳参加，安琦教授审阅了全书并提出了宝贵意见。教材的修订还得到了华东理工大学教务处、机械与动力工程学院的大力支持，彭建华教授在收集资料方面提供了许多帮助，在此一并谨致以衷心的感谢。

由于编者水平有限，难免存在不足之处，敬请读者批评指正。

编　者

2020 年 4 月

第一版前言

出版系列专业英语教材是许多院校师生多年来共同的愿望。为满足面向 21 世纪高等教育改革的需要，化学工业出版社及时与原化工部教育主管部门和全国化工类相关专业教学指导委员会协商，组织全国十余所院校成立了大学英语专业阅读教材编委会。在经过必要的调研后，根据学校需求，编委会优先从各高校教学（交流）讲义中确定选题，同时组织力量开展编审工作。本套教材涉及的专业主要包括机械工程、化学工程与工艺、信息工程、工业自动化、应用化学及精细化工、生物工程、环境工程、材料科学与工程、制药工程等。

根据“全国部分高校化工类及相关专业大学英语专业阅读教材编审委员会”的要求和安排编写的《机械设计制造及其自动化专业英语》教材，可供机械工程及相关专业本科生使用，也可作为同等程度的专业技术人员的自学教材。

本教材共分为三部分（Part），30 个单元（Unit）。第一部分为机械设计与制造的基本知识；第二部分为自动控制的基本知识；第三部分为提高部分，主要介绍现代先进制造技术。每个单元由主课文、主课文词汇表、课文注释、练习作业、阅读材料和阅读材料词汇表组成。书后还附有词汇总表。

本教材的内容覆盖了机械设计制造及其自动化专业的基本内容。材料均选自近年原版英文著作、教材、科技报告和专业期刊，并兼顾多种体裁以及英美的不同文风。各单元之间，既有一定的内在联系，又独立成章，可根据不同学时数灵活选用。

本教材由华东理工大学马玉录和大连理工大学刘东学主编。第一部分由大连理工大学刘东学、谢洪勇、李惠荣、李惠玲、银建中编写，第二部分和第三部分由华东理工大学马玉录、周邵萍、李琳、洪瑛编写。上海交通大学的蔡建国教授审阅了全书，并提出了宝贵意见。在本书的编写过程中得到了大学英语专业阅读教材编委会、华东理工大学教务处和大连理工大学教务处的大力支持，华东理工大学研究生金彦、何晓薇、关建生、本科生郭永征在本书的录入过程中做了大量工作，在此一并谨致以衷心的感谢。

限于编者水平，难免存在不足之处，热诚希望使用本书的广大师生提出宝贵意见。

编　者

2001 年 5 月

第二版前言

出版系列专业英语教材是许多院校师生多年来共同的愿望。为满足面向21世纪高等教育改革的需要，化学工业出版社及时与原化工部教育主管部门和全国化工类相关专业教学指导委员会协商，组织全国十余所院校成立了大学英语专业阅读教材编委会。在经过必要的调研后，根据学校需求，编委会优先从各高校教学（交流）讲义中确定选题，同时组织力量开展编审工作。本套教材涉及的专业主要包括机械工程、化学工程与工艺、信息工程、工业自动化、应用化学及精细化工、生物工程、环境工程、材料科学与工程、制药工程等。

根据“全国部分高校化工类及相关专业大学英语专业阅读教材编审委员会”的要求和安排编写的《机械设计制造及其自动化专业英语》教材，可供机械工程及相关专业本科生使用，也可作为同等程度的专业技术人员的自学教材。

本教材共分为三部分（Part），30个单元（Unit）。第一部分为机械设计与制造的基本知识；第二部分为自动控制的基本知识；第三部分为提高部分，主要介绍现代先进制造技术。每个单元由主课文、主课文词汇表、课文注释、练习作业、阅读材料和阅读材料词汇表组成。书后还附有词汇总表。

本教材的内容覆盖了机械设计制造及其自动化专业的基本内容。材料均选自近年原版英文著作、教材、科技报告和专业期刊，并兼顾多种体裁以及英美的不同文风。各单元之间，既有一定的内在联系，又独立成章，可根据不同学时数灵活选用。

《机械设计制造及其自动化专业英语》教材自2001年5月出版以来，一直得到同行以及广大师生的支持和爱护。几年来，我们不断征求大家对本教材的意见和建议，根据调研情况以及我们的教学实践，修订了本教材。

机械设计制造及其自动化专业英语（第二版）对部分内容作了更换与调整，增加了如模具设计与制造、过程控制系统状态监测、精益制造等内容，使教材适应性更广，也更加通俗易懂。

本教材的修订工作主要由华东理工大学马玉录、周邵萍、李琳、洪瑛参加。上海交通大学蔡建国教授在教材的编写过程中给予了许多指导，提出了宝贵意见并审阅了全书，在此一并致以衷心的感谢。

限于编者水平，难免存在不足之处，我们热诚希望使用本书的广大师生提出宝贵意见。

编　者

2008年10月

第三版前言

《机械设计制造及其自动化专业英语》（第二版）于 2009 年 1 月出版以来，已先后重印了多次，受到了本专业和相关专业教师和学生以及专业技术人员的支持和爱护。本教材还先后获得中国石油和化学工业联合会 2010 年度中国石油和化学工业优秀出版物奖（教材奖）一等奖和 2011 年上海普通高校优秀教材奖二等奖。

根据我们的教学实践和听取其他高校在使用本教材过程中提出的宝贵意见和建议，在前一版的基础上对全书进行了修订和补充。第三版基本结构和格式上与第二版保持一致，根据教学需要更换和调整了部分内容，增加了如材料力学绪论、阀门、热力学应用、控制模式等内容，保留的内容尽可能使用最新版本的内容。这样，使教材的内容具有更强的时效性和适用性。

本教材的修订工作主要由华东理工大学马玉录、周邵萍、李琳、洪瑛参加。上海交通大学蔡建国教授在教材的编写过程中给予了许多指导，提出了宝贵意见并审阅了全书，在此一并谨致以衷心的感谢。

由于编者水平有限，难免存在不足之处，敬请读者批评指正。

编　者

2014 年 10 月

Contents

PART Ⅰ

FUNDAMENTALS OF MACHINE DESIGN & MANUFACTURING

Unit 1 • Metals

The use of metals has always been a key factor in the development of the social systems of man. Of the roughly 100 basic elements of which all matter is composed, about half are classified as metals. The distinction between a metal and a nonmetal is not always clear cut. The most basic definition centers around the type of bonding existing between the atoms of the element, and around the characteristics of certain of the electrons associated with these atoms①. In a more practical way, however, a metal can be defined as an element which has a particular package of properties.

Metals are crystalline when in the solid state and, with few exceptions (e. g., mercury), are solid at ambient temperatures. They are good conductors of heat and electricity and are opaque to light. They usually have a comparatively high density. Many metals are ductile——that is, their shape can be changed permanently by the application of a force without breaking. The forces required to cause this deformation and those required finally to break or fracture a metal are comparatively high, although, the fracture force is not nearly as high as would be expected from simple considerations of the forces required to tear apart the atoms of the metal②.

One of the more significant of these characteristics from our point of view is that of crystallinity. A crystalline solid is one in which the constituent atoms are located in a regular three-dimensional array as if they were located at the corners of the squares of a three-dimensional chessboard③. The spacing of the atoms in the array is of the same order as the size of the atoms, the actual spacing being a characteristic of the particular metal. The directions of the axes of the array define the orientation of the crystal in space. The metals commonly used in engineering practice are composed of a large number of such crystals, called grains. In the most general case, the crystals of the various grains are randomly oriented in space. The grains are everywhere in intimate contact with one another and joined together on an atomic scale. The region at which they join is known as a grain boundary.

An absolutely pure metal (i. e., one composed of only one type of atom) has never been produced. Engineers would not be particularly interested in such a metal even if it were

to be produced, because it would be soft and weak. The metals used commercially inevitably contain small amounts of one or more foreign elements, either metallic or nonmetallic. These foreign elements may be detrimental, they may be beneficial, or they may have no influence at all on a particular property. If disadvantageous, the foreign elements tend to be known as impurities. If advantageous, they tend to be known as alloying elements. Alloying elements are commonly added deliberately even in substantial amounts in engineering materials. The result is known as an alloy.

The distinction between the descriptors "metal" and "alloy" is not clear cut. The term "metal" may be used to encompass both a commercially pure metal and its alloys. Perhaps it can be said that the more deliberately an alloying addition has been made and the larger the amount of the addition, the more likely it is that the product will specifically be called an alloy. In any event, the chemical composition of a metal or an alloy must be known and controlled within certain limits if consistent performance is to be achieved in service. Thus chemical composition have to be taken into account when developing an understanding of the factors which determine the properties of metals and their alloys.

Of the 50 or so metallic elements, only a few are produced and used in large quantities in engineering practice. The most important by far is iron, on which are based the ubiquitous steels and cast irons (basically alloys of iron and carbon). They account for about 98% by weight of all metals produced. Next in importance for structural uses (that is, for structures that are expected to carry loads) are aluminum, copper, nickel, and titanium. Aluminum accounts for about 0.8% by weight of all metals produced, and copper about 0.7%, leaving only 0.5% for all other metals. As might be expected, the remainder are all used in rather special applications. For example, nickel alloys are used principally in corrosion-and heat-resistant applications, while titanium is used extensively in the aerospace industry because its alloys have good combinations of high strength and low density. Both nickel and titanium are used in high-cost, high-quality applications, and, indeed, it is their high cost that tends to restrict their application.

We cannot discuss these more esoteric properties here. Suffice it to say that a whole complex of properties in addition to structural strength is required of an alloy before it will be accepted into, and survive in, engineering practice④. It may, for example, have to be strong and yet have reasonable corrosion resistance; it may have to be able to be fabricated by a particular process such as deep drawing, machining, or welding; it may have to be readily recyclable; and its cost and availability may be of critical importance.

Selected from "Metals Engineering A Technical Guide", Leonard E. Samuels, Carnes Publication Services, Inc., 1988.

New Words and Expressions

1. **nonmetal** [nɔn'metl] *n.* 非金属
2. **crystalline** ['kristəlain] *a.* 结晶性的，晶状的
3. **ambient** ['æmbiənt] *a.*; *n.* 周围的；周围环境
4. **ambient temperature** 室温，环境温度

5. opaque [əu'peik] *a*. 不透明的
6. ductile ['dʌktail] *a*. 延性的，易变形的，可塑的，韧性的
7. deformation [diːfɔː'meiʃən] *n*. 变形
8. crystallinity [kristə'liniti] *n*. (结) 晶性，结晶度
9. constituent [kən'stitjuənt] *a*.；*n*. 组成的，构成的；成分，组分
10. dimensional [dai'menʃənəl] *a*. 线 (维) 度的，…维的
11. orientation [ˌɔːrien'teiʃən] *n*. 定向，定位，排列方向
12. grain [grein] *n*. 颗粒，晶粒
13. grain boundary 晶界
14. ubiquitous [juː'bikwitəs] *a*. 处处存在的，普遍存在的
15. cast irons 铸铁
16. corrosion [kə'rəuʒən] *n*. 腐蚀
17. esoteric [ˌesəu'terik] *a*. 深奥的，奥秘的
18. fabricate ['fæbrikeit] *vt*. 制造加工

Notes

① 参考译文：最基本的定义是以围绕存在于元素原子间的键接类型以及与这些原子联系的电子某些特性为主。

这里 associated with 的意思是“与……有关系”。

② 参考译文：引起永久变形所需的力和最终使金属断裂所需的力相当大，尽管发生断裂所需的力没有像所预期的撕开金属原子所需的力那么大。

③ 参考译文：结晶固体是这样一种结构，组成它的原子定位在规则的三维排列中，仿佛位于三维棋盘的方格的角上。

此句的时态表达是因由 as if 引导的从句要求的虚拟语句所致。

④ 参考译文：在合金材料被采用和应用于工程实际之前，除需要掌握其结构强度外，还需知道它的综合性质就够了。

Suffice it to say that，意思为：“(只要) 说……就够了”。

Exercises

1. Answer the following questions according to the text.

   ① How many basic elements are classified as metal?

   ② What is a crystalline solid?

   ③ Which metallic elements are produced and used in large quantities in engineering practice?

   ④ What requirements are met before an alloy will survive in engineering practice?

2. Translate the 6th paragraph into Chinese.

3. Put the following into Chinese by reference to the text.

   aluminum　copper　nickel　titanium　structural strength　deep drawing

4. Put the following into English.

   定义　力　轴　非金属　结构　载荷　用途　性质

5. Translate the following sentences into English.

   ① 金属和非金属的差异一般很难界定。

   ② 即使绝对纯金属可以生产出来，工程师们对它并不特别感兴趣。

③ 在 50 种左右的金属元素里，工程实践中只有少数金属被大量生产和使用。

Reading Material 1

Stainless Steels

Stainless steels do not rust in the atmosphere as most other steels do. The term "stainless" implies a resistance to staining, rusting, and biting in the air, moist and polluted as it is, and generally defines a chromium content in excess of 11% but less than 30%. And the fact that the stuff is "steel" means that the base is iron.

Stainless steels have room-temperature yield strengths that range from 205 MPa (30 ksi) to more than 1,725 MPa (250 ksi). Operating temperatures around 750℃ (1,400 ℉) are common, and in some applications temperatures as high as 1090℃ (2,000 ℉) are reached. At the other extreme of temperature some stainless steels maintain their toughness down to temperatures approaching absolute zero.

With specific restrictions in certain types, the stainless steels can be shaped and fabricated in conventional ways. They can be produced and used in the as-cast condition; shapes can be produced by powder-metallurgy techniques; cast ingots can be rolled or forged (and this accounts for the greatest tonnage by far). The rolled product can be drawn, bent, extruded, or spun. Stainless steel can be further shaped by machining, and it can be joined by soldering, brazing, and welding. It can be used as an integral cladding on plain carbon or low alloy steels.

The generic term "stainless steel" covers scores of standard compositions as well as variations bearing company trade names and special alloys made for particular applications. Stainless steels vary in their composition from a fairly simple alloy of, essentially, iron with 11% chromium, to complex alloys that include 30% chromium, substantial quantities of nickel, and half a dozen other effective elements. At the high-chromium, high-nickel end of the range they merge into other groups of heat-resisting alloys, and one has to be arbitrary about a cutoff point. If the alloy content is so high that the iron content is about half, however, the alloy falls outside the stainless family. Even with these imposed restrictions on composition, the range is great, and naturally, the properties that affect fabrication and use vary enormously. It is obviously not enough to specify simply a "stainless steel".

The various specifying bodies categorize stainless steels according to chemical composition and other properties. For example, the American Iron and Steel Institute (AISI) lists more than 40 approved wrought stainless steel compositions; the American Society for Testing and Materials (ASTM) calls for specifications that may conform to AISI compositions but additionally require certain mechanical properties and dimensional tolerances; the Alloy Casting Institute (ACI) specifies compositions for cast stainless steels within the categories of corrosion-and heat-resisting alloys; the Society of Automotive Engineers (SAE) has

adopted AISI and ACI compositional specifications. Military specification MIL-HDBK-5 lists design values. In addition, manufacturers' specifications are used for special purposes or for proprietary alloys. Federal and military specifications and manufacturers' specifications are laid down for special purposes and sometimes acquire a general acceptance.

However, all the stainless steels, whatever specifications they conform to, can be conveniently classified into six major classes that represent three distinct types of alloy constitution, or structure. These classes are ferritic, martensitic, austenitic, manganese-substituted austenitic, duplex austenitic-ferritic, and precipitation-hardening.

Ferritic Stainless steel is so named because the crystal structure of the steel is the same as that of iron at room temperature. The alloys in the class are magnetic at room temperature and up to their Curie temperature [about 750 ℃ (1,400 ℉)]. Common alloys in the ferritic class contain between 11% and 29% chromium, no nickel, and very little carbon in the wrought condition. The 11% ferritic chromium steels, which provide fair corrosion resistance and good fabrication at low cost, have gained wide acceptance in automotive exhaust systems, containers, and other functional applications. The intermediate chromium alloys, with 16%~17% chromium, are used primarily as automotive trim and cooking utensils, always in light gages, their use somewhat restricted by welding problems. The high-chromium steels, with 18% to 29% chromium content, have been used increasingly in applications requiring a high resistance to oxidation and, especially, to corrosion. These alloys contain either aluminum or molybdenum and have a very low carbon content.

The high-temperature form of iron (between 910 ℃ and 1,400 ℃, or 1,670 ℉ and 2,550 ℉) is known as austenite (Strictly speaking the term austenite also implies carbon in solid solution). The structure is nonmagnetic and can be retained at room temperature by appropriate alloying. The most common austenite retainer is nickel. Hence, the traditional and familiar austenitic stainless steels have a composition that contains sufficient chromium to offer corrosion resistance, together with nickel to ensure austenite at room temperature and below. The basic austenitic composition is the familiar 18% chromium, 8% nickel alloy. Both chromium and nickel contents can be increased to improve corrosion resistance, and additional elements (most commonly molybdenum) can be added to further enhance corrosion resistance.

The justification for selecting stainless steel is corrosion and oxidation resistance. Stainless steels possess, however, other outstanding properties that in combination with corrosion resistance contribute to their selection. These are the ability to develop very high strength through heat treatment or cold working; weldability; formability; and in the case of austenitic steels, low magnetic permeability and outstanding cryogenic mechanical properties.

The choice of a material is not simply based on a single requirement, however, even though a specific condition (for example, corrosion service) may narrow the range of possibilities. For instance, in the choice of stainless steel for railroad cars, while corrosion resistance is one determining factor, strength is particularly significant. The higher price of stainless steel

compared with plain carbon steel is moderated by the fact that the stainless has about twice the allowable design strength. This not only cuts the amount of steel purchased, but by reducing the dead weight of the vehicle, raises the load that can be hauled. The same sort of reasoning is even more critical in aircraft and space vehicles.

But weight saving alone may be accomplished by other materials, for examples, the high-strength low-alloy steels in rolling stock and titanium alloys in aircraft. Thus, the selection of a material involves a careful appraisal of all service requirements as well as a consideration of the ways in which the required parts can be made. It would be foolish to select material on the basis of its predicted performance if the required shape could be produced only with such difficulty that cost skyrocketed.

The applicability of stainless steels may be limited by some specific factor, for example, an embrittlement problem or susceptibility to a particular corrosive environment. In general terms, the obvious limitations are:

① In chloride environments susceptibility to pitting or stress-corrosion cracking requires careful appraisal. One cannot blindly assume that a stainless steel of some sort will do. In fact, it is possible that no stainless will serve.

② The temperature of satisfactory operation depends on the load to be supported, the time of its application, and the atmosphere. However, to offer a round number for the sake of marking a limit, we suggest a maximum temperature of 870 ℃ (1,600 ℉). Common stainless steels can be used for short times above this temperature, or for extended periods if the load is only a few thousand pounds per square inch. But if the loads or the operating periods are great, then more exotic alloys are called for.

Selected from "Stainless Steel", R. A. Lula, American Society for Metals, 1986.

New Words and Expressions

1. as-cast [æz'kɑːst] *a.* 铸态的
2. powder-metallurgy ['paudəme'tælədʒi] *n.* 粉末冶金学
3. cast ingot [kɑːst iŋgət] *n.* 铸锭
4. roll [rəul] *v.* 轧制
5. tonnage ['tʌnidʒ] *n.* （总）吨位
6. extrude [eks'truːd] *v.* 挤压
7. spin [spin] *v.* 旋压
8. solder ['sɔːldə] *vt.* 钎焊
9. braze [breiz] *vt.* 铜焊
10. cladding ['klædiŋ] *n.* 包层，覆盖，（金属）覆层
11. wrought [rɔːt] *a.* 可锻的
12. American Iron and Steel Institute (AISI) 美国钢铁学会
13. American Society for Testing and Materials (ASTM) 美国材料试验学会
14. Alloy Casting Institute (ACI) 合金铸造学会
15. Society of Automotive Engineers (SAE) 美国汽车工程师学会

16. ferritic [fə'ritik] *a*. 铁素体的
17. martensitic [mɑː'tenzaitik] *a*. 马氏体的
18. austenitic [ˌɔːstə'nitik] *a*. 奥氏体的
19. oxidation [ɔksi'deiʃən] *n*. 氧化
20. cryogenic ['kraiədʒenik] *a*. 低温的，深冷的

Unit 2 • Selection of Construction Materials

There is not a great difference between "this" steel and "that" steel; all are very similar in mechanical properties. Selection must be made on factors such as hardenability, price, and availability, and not with the idea that "this" steel can do something no other can do because it contains 2 percent instead of 1 percent of a certain alloying element, or because it has a mysterious name. A tremendous range of properties is available in any steel after heat treatment; this is particularly true of alloy steels.

Considerations in Fabrication

The properties of the final part (hardness, strength, and machinability), rather than properties required by forging, govern the selection of material. The properties required for forging have very little relation to the final properties of the material; therefore, not much can be done to improve its forgeability. Higher-carbon steel is difficult to forge. Large grain size is best if subsequent heat treatment will refine the grain size.

Low-carbon, nickel-chromium steels are just about as plastic at high temperature under a single 520 ft • lb (1 ft • lb=1. 355, 82 J) blow as plain steels of similar carbon content. Nickel decreases forgability of medium-carbon steels, but has little effect on low-carbon steels. Chromium seems to harden steel at forging temperatures, but vanadium has no discernible effect; neither has the method of manufacture any effect on high-carbon steel.

Formability

The cold-formability of steel is a function of its tensile strength combined with ductility. The tensile strength and yield point must not be high or too much work will be required in bending; likewise, the steel must have sufficient ductility to flow to the required shape without cracking. The force required depends on the yield point, because deformation starts in the plastic range above the yield point of the steel. Work-hardening also occurs here, progressively stiffening the metal and causing difficulty, particularly in the low-carbon steels.

It is quite interesting in this connection to discover that deep draws can sometimes be made in one rapid operation that could not possibly be done leisurely in two or three①. If a draw is half made and then stopped, it may be necessary to anneal before proceeding, that is, if the piece is given time to work-harden. This may not be a scientific statement, but it is actually what seems to happen.

Internal Stresses

Cold forming is done above the yield point in the work-hardening range, so internal stresses can be built up easily. Evidence of this is the springback as the work leaves the forming operation and the warpage in any subsequent heat treatment. Even a simple washer might, by virtue of the internal stresses resulting from punching and then flattening, warp severely during heat treating②.

When doubt exists as to whether internal stresses will cause warpage, a piece can be checked by heating it to about 1,100 ℉ and then letting it cool. If there are internal stresses, the piece is likely to deform. Pieces that will warp severely while being heated have been seen, yet the heat-treater was expected to put them through and bring them out better than they were in the first place.

Welding

The maximum carbon content of plain carbon steel safe for welding without preheating or subsequent heat treatment is 0.30%. Higher-carbon steel is welded every day, but only with proper preheating. There are two important factors: (1) the amount of heats that is put in; (2) the rate at which it is removed.

Welding at a slower rate puts in more heat and heats a large volume of metal, so the cooling rate due to loss of heat to the base metal is decreased. A preheat will do the same thing. For example, SAE 4,150 steel, preheated to 600 ℉ or 800 ℉, can be welded readily. When the flame or arc is taken away from the weld, the cooling rate is not so great, owing to the higher temperature of the surrounding metal, and slower cooling results. Even the most rapid air-hardening steels are weldable if preheated and welded at a slow rate.

Machinability

Machinability means several things. To production men it generally means being able to remove metal at the fastest rate, leave the best possible finish, and obtain the longest possible tool life③. Machinability applies to the tool-work combination.

It is not determined by hardness alone, but by the toughness, microstructure, chemical composition, and tendency of a metal to harden under cold work. In the misleading expression "too hard to machine", the work "hard" is usually meant to be synonymous with "difficult". Many times a material is actually too soft to machine readily. Softness and toughness may cause the metal to tear and flow ahead of the cutting tool rather than cut cleanly. Metals that are inherently soft and tough are sometimes alloyed to improve their machinability at some sacrifice in ductility. Examples are use of lead in brass and of sulfur in steel.

Machinability is a term used to indicate the relative ease with which a material can be machined by sharp cutting tools in operations such as turning, drilling, milling, broaching, and reaming.

In the machining of metals, the metal being cut, the cutting tool, the coolant, the process and type of machine tool, and the cutting conditions all influence the results. By changing any one of these factors, different results will be obtained. The criterion upon which the ratings listed are based is the relative volume of various materials that may be removed by turning under fixed conditions to produce an arbitrary fixed amount of tool wear.

Selected from "Modern Manufacturing Process Engineering", Benjamin W. Niebel, *et al.*, McGraw-Hill Publishing Company, 1989.

New Words and Expressions

1. availability [əˌveiləˈbiliti] *n.* 可用性，有效性，可得性

2. fabrication [ˌfæbriˈkeiʃən] *n*. 制造
3. forgeability [ˈ fɔːdʒəˈbiliti] *n*. 可锻性
4. nickel [ˈnikl] *n*. 镍
5. chromium [ˈkroumjəm] *n*. 铬
6. vanadium [vəˈneidiəm] *n*. 钒
7. discernible [diˈsəːnəbl] *a*. 可辨别得出的，可看出的
8. ductility [dʌkˈtiliti] *n*. 延（展）性，韧性
9. cracking [ˈkrækiŋ] *n*. 开裂，裂纹，裂缝
10. work-harden 加工硬化，冷作硬化
11. anneal [əˈniːl] *n*.；*v*. 退火
12. warp [wɔːp] *v*. 翘曲，变形
13. preheat [priːˈhiːt] *v*. 预热
14. microstructure [ˈmaikrəuˈstrʌktʃə] *n*. 显微结构
15. mislead [misˈliːd] *vt*. 使……误解，误导
16. ream [riːm] *vt*. （用铰刀）铰孔
17. arbitrary [ˈɑːbitrəri] *a*. 任意的

Notes

① 参考译文：在这方面，相当有趣的是你将发现有时可通过一次快速加载完成深度拉伸，但以缓慢的方式两三次加载却不能实现。

in this connection 意思为在这方面。deep draws，深度拉伸。

② 参考译文：即使一个简单的垫圈，由于冲孔与随后的压平而导致内应力，也会在热处理时引起严重翘曲。

③ 参考译文：对于（机械加工）工人来说，可加工性通常意味着能够以最快的速度切削工件，获得最好的表面光洁度，并使刀具保持最长的使用寿命。

Exercises

1. After reading the text above，write a summary of it.
2. Answer the following questions according to the text.

① What basic concepts assist the production-design engineer in the selection of steel?

② What is the most important factor in selection of material when a series of manufacturing process is needed?

③ Which mechanical property or behavior is most important for the formability of the materials?

④ How can a piece of steel be checked to determine whether internal stresses will cause warpage?

3. Translate the 1st paragraph into Chinese.
4. Put the following into Chinese by reference to the text.

hardenability　machinability　cold drawn　steel sheet　percent reduction in area

endurance limit　rolled-steel shapes　corrosion resistance　rupture

5. Put the following into English.

低碳钢　高强度钢　热处理　屈服强度　机械性能　韧性　内应力

Reading Material 2

Polymer and Composites

Polymers and polymer composites are used in many different forms, ranging from synthetics through to structural composites in the construction industry and to the high technology composites of the aerospace and space satellite industries.

Plastics are generally considered to be a relatively recent development. In fact, they are members of the much larger family of polymers.

Polymers are the products of combining a large number of small molecular units called monomers by the chemical process known as polymerization (which is the process by which molecules or groups of atoms are joined together) to form long-chain molecules. Natural materials such as bitumen, rubber and cellulose have this type of structure. There are two main types of polymerization. In the first type, a substance consisting of a series of long-chain polymerized molecules, called thermoplastics, is produced. All the chains of the molecules are separate and can slide over one another. In the second type, the chains become cross-linked so that a solid material is produced which cannot be softened and which will not flow. Such solids are called thermosetting polymers. These two groups classify polymer materials.

Polymers are usually made in one of two polymerization processes. In condensation polymerization the linking of molecules creates by-products, usually water, nitrogen or hydrogen gas. In addition polymerization no by-products are created.

The final property of the plastics material will be determined by the type of additive mixed with the pure polymer. Additives may be simply fillers or extenders designed to reduce the quantity of polymers used or to toughen the product. Pigments or stabilizers can be added to provide colouring for the material or to reduce degradation, and plasticizers are incorporated to alter the characteristics of the polymers. The glossary outlines the purpose of the main additives used with polymers.

Thermoplastic polymers consist of linear molecules which are not interconnected. The chemical valency bond along the chain is extremely strong, but the forces of attraction between the adjacent chains are weak. Because of their unconnected chain structure, thermoplastics may be repeatedly softened and hardened by heating and cooling respectively; with each repeated cycle, however, the material tends to become more brittle.

The thermosetting polymer is formed by a chemical reaction. In the first stage, a substance consisting of a series of long-chain polymerized molecules, similar to those present in thermoplastics, is produced. In the second stage of the process, the chains become cross-linked; this reaction can take place either at room temperature or under the application of heat and pressure. The resultant material will not flow and cannot be softened by heating.

The mechanical properties of polymers can be greatly improved by using techniques from nature. Few natural materials consist of one substance only: most are a mixture of different components which, when combined, produce a material with enhanced properties. Bone, for example, achieves its lightness and strength by combining crystals of apatite with fibres of the protein collagen. This combination of two or more materials is known as a composite.

Polymers are often combined with fillers and/or fibres to improve their mechanical or physical properties. The fillers usually consist of wood flour, china clay, quartz powder or other powdered minerals. The filler is incorporated not only to improve the physical property of the composite but also sometimes to reduce the polymer content and hence the cost.

The use of fibres in combination with polymers enhances the latter's mechanical properties for structural use. Most fibres can be used, provided there is chemical compatibility between the two parts.

Both thermoplastics and thermosetting polymers can be combined with fibre reinforcement. In the past, thermoplastics have been combined mainly with short fibres but these composites were unable to take full advantage of the fibre strength. However, users have taken advantage of this system's toughness property by manufacturing a range of injection mouldings. Conversely, the thermosetting composite system tends to be brittle, because of the cross-linking of the matrix. Long fibres have been utilized in the composite, to take full advantage of the strength of the fibre. This system has found application in large panels and pressure pipes and containers.

Currently, however, the traditional market areas of the thermoplastics and thermosetting polymer composites are less well defined and both long and short fibres are being used with thermoplastics and thermosetting matrices.

Selected from "Polymers and Polymer Composites in Construction", L. C. Hollaway Thomas, Telford Ltd, London, 1990.

New Words and Expressions

1. polymer ['pɔlimə] *n*. 聚合物，聚合材料
2. composite [kəm'pɔzit] *n*. 合成，复合，复合材料
3. monomer ['mɔnəmə] *n*. 单（分子物）体，单基物
4. polymerization [pɔliməraiˈzeiʃən] *n*. 聚合作用，聚合反应
5. thermoplastic ['θəːməu'plæstik] *a*.; *n*. 热塑性的；热塑性塑料
6. thermosetting [ˌθəːməu'setiŋ] *a*. 热固（凝）的
7. thermosetting polymer 热固性塑料
8. additive ['æditiv] *n*.; *a*. 添加剂；增加的
9. filler ['filə] *n*. 填充物，填料
10. extender [iks'tendə] *n*. 填充剂，补充料
11. pigment ['pigmənt] *n*. 染料，色素
12. stabilizer ['steibilaizə] *n*. 稳定剂
13. degradation [degrə'deiʃən] *n*. 退化，降低，劣化
14. plasticizer ['plæstisaizə] *n*. 增塑剂，柔韧剂
15. adjacent [ə'dʒeisənt] *a*. 相邻的，毗连的

16. brittle ['britl] *a*. 易碎的，脆性的
17. enhance [in'hɑːns] *v*. 提高，增强
18. quartz [kwɔːts] *n*. 石英
19. compatibility [kəmˌpætə'biliti] *n*. 相容性，可混性
20. moulding ['məuldiŋ] *n*. 模塑（法），造型（法）
21. injection moulding （塑料）注射成型
22. matrix ['meitriks] *n*. 基体，基质，矩阵

Unit 3 • Mechanical Properties of Materials

The material properties can be classified into three major headings: (ⅰ) Physical, (ⅱ) Chemical, (ⅲ) Mechanical.

Physical Properties

Density or specific gravity, moisture content, etc., can be classified under this category.

Chemical Properties

Many chemical properties come under this category. These include acidity or alkalinity, reactivity and corrosion. The most important of these is corrosion which can be explained in layman's terms as the resistance of the material to decay while in continuous use in a particular atmosphere.

Mechanical Properties

Mechanical properties include the strength properties like tensile, compression, shear, torsion, impact, fatigue and creep. The tensile strength of a material is obtained by dividing the maximum load, which the specimen bears by the area of cross-section of the specimen①.

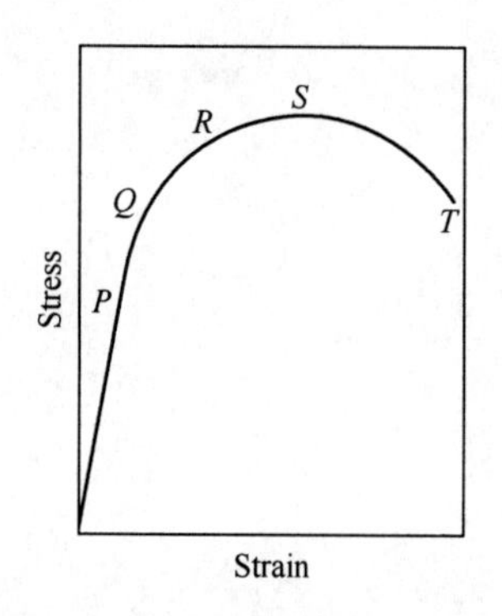

Fig. 3. 1 Stress-Strain curve

This is a curve plotted between the stress along the *Y*-axis (ordinate) and the strain along the *X*-axis (abscissa) in a tensile test. A material tends to change or changes its dimensions when it is loaded, depending upon the magnitude of the load. When the load is removed it can be seen that the deformation disappears. For many materials this occurs up to a certain value of the stress called the elastic limit σ_e②. This is depicted by the straight line relationship and a small deviation thereafter, in the stress-strain curve (Fig. 3. 1).

Within the elastic range, the limiting value of the stress up to which the stress and strain are proportional, is called the limit of proportionality σ_p. In this region, the metal obeys Hooke's law, which states that the stress is proportional to strain in the elastic range of loading (the material completely regains its original dimensions after the load is removed). In the actual plotting of the curve, the proportionality limit is obtained at a slightly lower value of the load than the elastic limit. This may be attributed to the time-lag in the regaining of the original dimensions of the material. This effect is very frequently noticed in some non-ferrous metals.

While iron and nickel exhibit clear ranges of elasticity, copper, zinc, tin, etc, are found to be imperfectly elastic even at relatively low values of stresses③. Actually the elastic limit is distinguishable from the proportionality limit more clearly depending upon the sensitivity of the measuring instrument.

When the load is increased beyond the elastic limit, plastic deformation starts. Simulta-

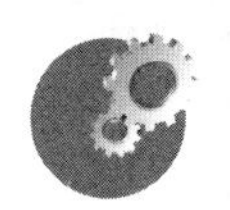

neously the specimen gets work-hardened. A point is reached when the deformation starts to occur more rapidly than the increasing load. This point is called the yield point Q. The metal which was resisting the load till then, starts to deform somewhat rapidly, i. e. , yield. The yield stress is called yield limit σ_y.

The elongation of the specimen continues from Q to S and then to T. The stress-strain relation in this plastic flow period is indicated by the portion $QRST$ of the curve. At T the specimen breaks, and this load is called the breaking load. The value of the maximum load S divided by the original cross-sectional area of the specimen is referred to as the ultimate tensile strength of the metal or simply the tensile strength σ_u.

Logically speaking, once the elastic limit is exceeded, the metal should start to yield, and finally break, without any increase in the value of stress. But the curve records an increased stress even after the elastic limit is exceeded. Two reasons can be given for this behavior:

① the strain hardening of the material;

② the diminishing cross-sectional area of the specimen, suffered on account of the plastic deformation④.

The more plastic deformation the metal undergoes, the harder it becomes, due to work-hardening. The more the metal gets elongated the more its diameter (and hence, cross-sectional area) is decreased. This continues until the point S is reached.

After S, the rate at which the reduction in area takes place, exceeds the rate at which the stress increases. Strain becomes so high that the reduction in area begins to produce a localized effect at some point. This is called necking.

Reduction in cross-sectional area takes place very rapidly; so rapidly that the load value actually drops. This is indicated by ST. Failure occurs at this point T.

Then percentage elongation δ and reduction in area Ψ indicate the ductility or plasticity of the material:

$$\delta=\frac{L-L_0}{L_0}\times 100\%$$

$$\Psi=\frac{A_0-A}{A_0}\times 100\%$$

Where L_0 and L are the original and the final length of the specimen; A_0 and A are the original and the final cross-section area.

Selected from "Testing of Metallic Materials", Prof. A. U. K. SURYANRAYANA, Prentice Hall of Lndia Private Limited, New Delhi-110001, 1979.

New Words and Expressions

1. alkalinity [ˌælkə'liniti] *n.* 碱性，碱度
2. tensile ['tensail] *a.* 受拉的，拉伸的
3. compression [kəm'preʃən] *n.* 压缩，压力
4. shear [ʃiə] *n.* 剪切，切应变
5. torsion ['tɔːʃən] *n.* 扭转，转矩

6. impact [ˈimpækt] *n.* 冲击，冲力，影响
7. fatigue [fəˈtiːg] *n.* 疲劳
8. creep [kriːp] *n.* 蠕变
9. specimen [ˈspesimən] *n.* 试件，试样
10. cross-section 横截面
11. ordinate [ˈɔːdineit] *n.* 纵坐标
12. abscissa [əbˈsisə] *n.* 横坐标
13. deviation [ˌdiːviˈeiʃən] *n.* 偏离，偏移
14. time-lag 延时，落后，时滞
15. yield [jiːld] *n.*；*v.* 屈服
16. elongation [ˌiːlɔŋˈgeiʃən] *n* 延伸率，伸长
17. diminish [diˈminiʃ] *n.* 减小，缩小
18. necking [ˈnekiŋ] *n.* 颈缩，形成细颈现象

Notes

① 参考译文：材料的拉伸强度可通过用试件的横截面面积除试件承受的最大载荷得到。句中 divide…by（…），用（……）除……。

② 参考译文：对许多材料而言，在达到称为弹性极限的一定应力值 σ_e 之前，一直表现为这样的行为。

up to 一直到。

③ 参考译文：铁和镍有明显的弹性范围，而铜，锌，锡等，即使在相对低的应力下也表现为不完全弹性。

while 这里表示同时存在的两种情况的对比，引出并列分句。

④ 参考译文：出现该现象可以归结为两个原因：（ⅰ）材料的应变硬化，（ⅱ）由于塑性变形，使试件横截面面积逐渐缩小。

on account of …，因为，为了……缘故。

Exercises

1. After reading the text above summarize the main ideas of it in oral English.
2. Answer the following questions according to the text.

   ① What does Hooke's law state?

   ② Try to explain what is the yield limit of a material.

   ③ Why does the stress-strain curve record an increased stress after the elastic limit is exceeded?

   ④ What kind of behavior is called necking?
3. Translate the 6th paragraph into Chinese.
4. Put the following into Chinese by reference to the text.

   non-ferrous　　stress-strain curve　　yield point　　percentage elongation

   necking　　sensitivity
5. Put the following into English.

   应变硬化　　横截面　　断面收缩率　　比例极限

   屈服极限　　延性　　机械性质　　用……除……
6. Translate the following sentences into English.

① 当承载时，材料将发生变形，变形量取决于载荷的大小。

② 在弹性范围内，载荷被卸除后，试件的变形消失。

③ 逻辑上，一旦超过弹性极限，金属将开始屈服，并且在不增加应力值的情况下，也会最终断裂。

Reading Material 3

Introduction to Mechanics of Materials

Mechanics of materials is a branch of applied mechanics that deals with the behavior of solid bodies subjected to various types of loading. Other names for this field of study are strength of materials and mechanics of deformable bodies. The solid bodies considered in this book include bars with axial loads, shafts in torsion, beams in bending, and columns in compression.

The principal objective of mechanics of materials is to determine the stresses, strains, and displacements in structures and their components due to the loads acting on them. If we can find these quantities for all values of the loads up to the loads that cause failure, we will have a complete picture of the mechanical behavior of these structures. An understanding of mechanical behavior is essential for the safe design of all types of structures, whether airplanes and antennas, buildings and bridges, machines and motors, or ships and spacecraft. That is why mechanics of materials is a basic subject in so many engineering fields. Statics and dynamics are also essential, but those subjects deal primarily with the forces and motions associated with particles and rigid bodies. In mechanics of materials we go one step further by examining the stresses and strains inside real bodies, that is, bodies of finite dimensions that deform under loads. To determine the stresses and strains, we use the physical properties of the materials as well as numerous theoretical laws and concepts.

Theoretical analyses and experimental results have equally important roles in mechanics of materials. Often we use theories to derive formulas and equations for predicting mechanical behavior, but these expressions cannot be used in practical design unless the physical properties of the materials are known. Such properties are available only after careful experiments have been carried out in the laboratory. Furthermore, many practical problems are not amenable to theoretical analysis alone, and in such cases physical testing is a necessity.

The historical development of mechanics of materials is a fascinating blend of both theory and experiment—theory has pointed the way to useful results in some instances, and experiment has done so in others. Such famous persons as Leonardo da Vinci (1452-1519) and Galileo Galilei (1564-1642) performed experiments to determine the strength of wires, bars, and beams, although they did not develop adequate theories (by today's standards) to explain their test results. By contrast, the famous mathematician Leonhard Euler (1707-1783) developed the mathematical theory of columns and calculated the critical load of a column in 1744, long before any experimental evidence existed to show the significance of his results. Without appropriate tests to back up his theories, Euler's results remained unused for over a hundred

years, although today they are the basis for the design and analysis of most columns.

Normal Stress

The most fundamental concepts in mechanics of materials are stress and strain. These concepts can be illustrated in their most elementary form by considering a cylindrical bar subjected to axial forces. A cylindrical bar is a straight structural member having constant cross section throughout its length, and an axial force is a load directed along the axis of the member, resulting in either tension or compression in the bar. Examples are the tow bar and landing gear strut of an airplane, the members of a bridge truss, connecting rods in automobile engines, spokes of bicycle wheels, columns in buildings.

For discussion purposes, we will consider a tow bar of an airplane and isolate a segment of it as a free body [Fig. 3. 2(a)]. When drawing this free-body diagram, we disregard the weight of the bar itself and assume that the only active forces are the axial forces P at the ends. Next we consider two views of the bar, the first showing the bar before the loads are applied [Fig. 3. 2(b)] and the second showing it after the loads are applied [Fig. 3. 2(c)]. Note that the original length of the bar is denoted by the letter L, and the increase in length is denoted by the Greek letter δ (delta).

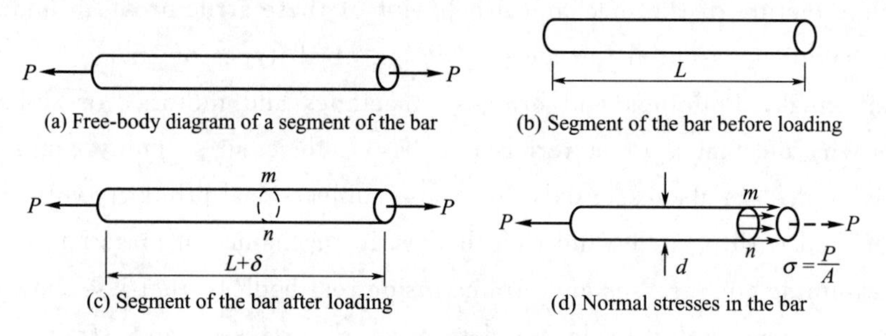

Fig. 3. 2 Cylindrical bar in tension

The internal stresses in the bar are exposed if we make an imaginary cut through the bar at section *mn* [Fig. 3. 2(c)]. Because this section is taken perpendicular to the longitudinal axis of the bar it is called a cross section. We now isolate the part of the bar to the left of cross section *mn* as a free body [Fig. 3. 2(d)]. At the right-hand end of this free body (section *mn*) we show the action of the removed part of the bar (that is, the part to the right of section *mn*) upon the part that remains. This action consists of a continuously distributed force acting over the entire cross section. The intensity of the force (that is, the force per unit area) is called the stress and is denoted by the Greek letter σ (sigma). Thus, the axial force P acting at the cross section is the resultant of the continuously distributed stresses. [The resultant force is shown with a dashed line in Fig. 3. 2(d)]

Assuming that the stresses are uniformly distributed over cross section *mn* [Fig. 3. 2 (d)], we see that their resultant must be equal to the intensity σ times the cross-sectional area A of the bar. Therefore, we obtain the following expression for the magnitude of the stresses:

$$\sigma=\frac{P}{A}$$

This equation gives the intensity of uniform stress in an axially loaded, cylindrical bar of arbitrary cross-sectional shape. When the bar is stretched by the forces P, the stresses are tensile stresses; if the forces are reversed in direction, causing the bar to be compressed, we obtain compressive stresses. Inasmuch as the stresses act in a direction perpendicular to the cut surface, they are called normal stresses. Thus, normal stresses may be either tensile or compressive.

When a sign convention for normal stresses is required, it is customary to define tensile stresses as positive and compressive stresses as negative.

Normal Strain

As already observed, a straight bar will change in length when loaded axially, becoming longer when in tension and shorter when in compression. For instance, consider again the cylindrical bar of Fig. 3. 2. The elongation δ of this bar [Fig. 3. 2(c)] is the cumulative result of the stretching of all elements of the material throughout the volume of the bar. Let us assume that the material is the same everywhere in the bar. Then, if we consider half of the bar (length $L/2$), it will have an elongation equal to $\delta/2$, and if we consider one-fourth of the bar, it will have an elongation equal to $\delta/4$. Similarly, a unit length of the bar will have an elongation equal to $1/L$ times the total elongation δ. By this process we arrive at the concept of elongation per unit length, or strain, denoted by the Greek letter ε (epsilon) and given by the equation

$$\varepsilon=\frac{\delta}{L}$$

If the bar is in tension, the strain is called a tensile strain, representing an elongation or stretching of the material. If the bar is in compression, the strain is a compressive strain and the bar shortens. Tensile strain is usually taken as positive and compressive strain as negative. The strain ε is called a normal strain because it is associated with normal stresses.

Selected from: "Mechanics of Materials (fourth edition)", James M. Gere, Stephen P. Timoshenko, PWS PUBLISHING COMPANY, 1997

New Words and Expressions

1. mechanics [mi'kæniks] *n.* 力学，机械 [构] 学；机构，结构
2. strength [streŋθ] *n.* 强度 (极限)，浓度，力 (量，气)
3. deformable body 可变形物体
4. axial ['æksiəl] *a.* 轴向的，轴线
5. load [ləud] *n.*; *v.* 载 [负] 荷，(荷，负) 载，加载
6. shaft [ʃaːft] *n.* 传动轴，旋转轴
7. beam [biːm] *n.* (横，天平) 梁；光 [射] 线
8. bend [bend] *v.* (使) 弯 [挠，折] 曲
   n. 弯 (管，头)，弯曲 (处)
9. stress [stres] *n.* 应力
10. strain [strein] *n.* 应变
11. displacement [dis'pleismənt] *n.* 位移，平移，偏移

12. failure [ˈfeiljə] *n*. 失效，破坏
13. statics [ˈstætiks] *n*. 静力学，静（止状）态
14. dynamics [daiˈnæmiks] *n*. 动力学，动态（特性）
15. particle [ˈpaːtikəl] *n*. [数，物] 质点，粒子
16. rigid body 刚体，刚性体
17. amenable [əˈmiːnəbl] *a*. 可处理的；经得起检验的
18. landing gear 起落架
19. strut [strʌt] *n*. 支柱；抗压构件
20. connecting rod 连杆，活塞杆
21. resultant [riˈzʌltənt] *n*.；*a*. 合力；合成的，总的

Unit 4 ● Application of Intensive Quenching Technology for Steel Parts

The rapid part cooling within the martensite transformation range can lead to distortion and quench cracks①. It is axiomatic that accelerated cooling results in high thermal and structural stresses. For years, heat treaters endeavoured to develop quench methods and quenchants to slow the rate of cooling in the martensite range, such as polymer, oil, hot oil, martemper salts and even interrupted quenches.

Counterintuitive to these "safe" quench practices are the "intensive" or "shell" quenching techniques. Intensive quenching may be defined as a very rapid, but uniform cooling, usually in a water/polymer bath with a high agitation rate. A high-pressure spray or water-mist quench, with very high evaporative cooling rates, also qualifies as an "intensive quench" and the spray or mist must be uniform over all part surfaces.

The reason why the intensive quenching process minimizes cracking and distortion may be explained by the following simplified model. Imagine a steel part with a varying thickness (Fig. 4. 1). During conventional quenching, the martensite forms first in the thinner section of the part since this section cools faster and reaches the martensite range earlier than the thicker one [Fig. 4. 1 (a)] . Because the martensite specific volume is greater than the specific volume of the remaining austenite, residual structural stresses result②. These stresses cause distortion and possible part cracking.

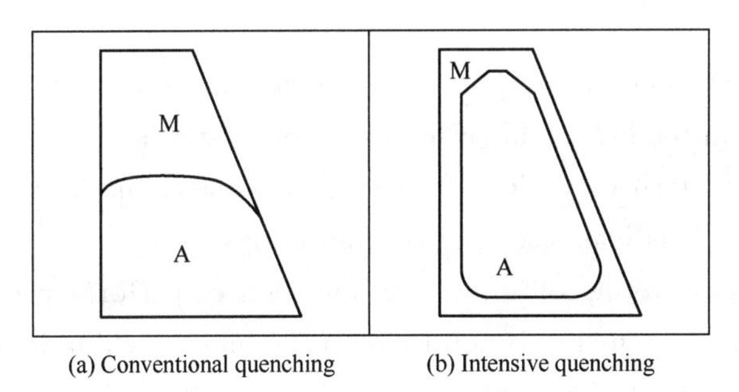

Fig. 4. 1 Schematic of martensite formation during conventional quenching and intensive quenching

A=austenite; M=martensite

Now imagine that the same steel part is cooled very rapidly. In this instance, the martensite forms simultaneously over the entire part surface creating a hardened shell [Fig. 4. 1 (b)] . This uniform hardened shell, with its high compressive stresses, contains the still-austenitic core during its transformation, netting higher surface hardness, deeper hardened layer, lower distortion, and lower probability of cracking. A key element of the intensive quenching process is rapid and uniform cooling interrupted when the compressive stresses on the surface are at the maximum value. It has been demonstrated that intensive quenching provides the following benefits:

① elimination of cracking;

② minimum distortion;

③ high compressive surface stresses (similar to those developed in peening);

④ improved mechanical properties (yield and ultimate strength, wear resistance, depth of hardness, etc.);

⑤ longer part life;

⑥ usage of less alloy steels (enhanced value engineering);

⑦ elimination of hazardous oil and hot salt quenchants with environmentally-friendly water or polymer/water solutions.

A computer model of the intensive quenching process has also been developed and validated, the model and its application has proven intensive quenching successful in the hardening of such steel parts as:

① auto-parts (semi-axles, spherical journals, springs);

② bearing rings;

③ fasteners (bolts, nuts, washers, etc.);

④ tools (dies, punches, etc.);

⑤ mining equipment parts;

⑥ large steel parts for machine building and power industries.

The intensive quenching process will be commercialized for two common methods of steel hardening: quenching single part and batch or multiple-part quenching. The commercialization of the technology includes the following steps:

① Computer simulations to develop the optimum time/temperature conditions for the specified steel parts;

② Verification of these conditions by performing experiments in the experimental quenching system or in the industrial-scale quenching system;

③ Additional statistical validation of the intensive quenching process over time and multiple quenches with various load sizes and configurations.

The simultaneous lowering of costs, optimization of performance and minimization of distortion of steel parts is a goal common to all in the metalworking industry. Demonstration of the intensive quenching technology presents many opportunities for part designers. The technology offers lower-cost alternatives to complement a wide array of alloy and micro-alloyed steels.

Lower distortion of intensive quenching offers the promise of eliminating many post-hardening, straightening or press-quenching operations on steel parts③. The additional benefits from replacing hazardous quenchants, with more environmentally-benign water and water/ polymer quenchants, are added incentives for the wide application of intensive quenching technology.

Selected from "Practical Application of Intensive Quenching Technology for Steel Parts", M. A. Aronov, *et al.*, Heat Treatment of Metals, No. 1, 2000.

New Words and Expressions

1. martensite [ˈmɑːtənzait] *n*. [冶] 马氏体
2. distortion [disˈtɔːʃən] *n*. 变形，畸变
3. quench [kwentʃ] *vt*. 把……淬火，使骤冷，使淬硬
4. axiomatic [æksiəˈmætik] *a*. 公理的，自明的，自然的
5. quenchant [ˈkwentʃənt] *n*. 淬火剂
6. martemper [ˈmɑːtempə] *n*. [冶] 等温淬火，热浴淬火，间歇淬火，马氏体回火
7. shell [ʃel] *n*. 壳，管壳，外壳；套管，罩
8. agitation [ædʒiˈteiʃən] *n*. (液体的) 搅动，摇动
9. mist [mist] *n*. 烟雾，薄雾，模糊
10. netting [ˈnetiŋ] *n*. 网，网状物，结网
11. austenite [ˈɔːstəˌnait] *n*. [冶] 奥氏体
12. transformation [ˌtrænsfəˈmeiʃən] *n*. 变形，转化，转换
13. peening [ˈpiːniŋ] *v*. [冶] 锤击硬化，喷丸硬化
14. validate [ˈvælideit] *vt*. 使生效，证实
15. spring [spriŋ] *n*. 弹簧，弹性，弹力
16. fastener [ˈfɑːsnə] *n*. 紧固件，接合件
17. washer [ˈwɔʃə] *n*. 垫圈，洗衣机，洗碗机
18. commercialization [kəˌməːʃəlaiˈzeiʃən] *n*. 商业化，商品化
19. performance [pəˈfɔːməns] *n*. 性能，特性；行为，操作，工况，绩效
20. benign [biˈnain] *a*. 有益于健康的，(气候) 温和的，良好的

Notes

① 参考译文：在马氏体转变区，零件的快速冷却可能导致零件产生变形与淬火裂纹。

② 参考译文：因为马氏体比容比剩余奥氏体比容大，导致产生残余应力。

③ 参考译文：强化淬火的较小变形，允许取消对钢制件进行后置硬化、校直或加压淬火等许多工序。

Exercises

1. After reading the text above, summarize the main ideas of it in oral English.
2. Answer the following questions according to the text.
   ① Why the intensive quenching process can play a role in minimizing cracking and distortion?
   ② What benefits can intensive quenching provide?
   ③ If the intensive quenching is commercialized, which steps should be taken?
   ④ Which promise has the lower distortion of intensive quenching offered on steel parts?
3. Translate the 1st and 2nd paragraphs into Chinese.
4. Put the following into Chinese by reference to the text.

   minimum distortion　yield and ultimate strength　wear resistance　optimization of performance
   intensive quenching　structural stress　bearing ring　compressive stress
5. Translate the following sentences into English.
   ① 在实现强化淬火工艺的工业化生产以前，对于一定的钢件，必须通过计算机模拟确定最佳的淬火

时间/温度条件。

② 在金属加工工业中，一个共同的目标就是降低成本，实现产品性能优化，减小钢件的变形量。

Reading Material 4

An Overview of Advances in Vacuum Heat Treatment

Heat Treatment in Vacuum Furnaces

Advances in vacuum heat treatment process and equipment have concentrated on faster and more uniform heating and, in the past few years especially, on enhancing the gas quenching effect. Emphasis has also been placed on the development of low-pressure carburizing, using hydrocarbon gases in vacuum furnaces with or without plasma assistance.

Different furnace designs have been developed for various applications ranging from single-chamber furnaces and double-or triple-chamber furnaces to integrated continuous vacuum furnace lines.

Heating in Vacuum Furnaces

Heating in pure vacuum relies solely on heat transfer by radiation. This effect is especially slow with dense loads consisting of a large number of components, where those in the center of the charge are shielded from the energy transfer. However, if gas can be circulated through the load during the heating phase, convective heat transfer will increase heating speed remarkably and reduce the heat-up time, especially in the low-temperature range (up to about 850 ℃).

Thus, in the late 1980's, vacuum hardening furnaces were developed with an incorporated dual-heating device. This consists of a convective heating system, working up to a temperature of the order of 850 ℃, and a pure radiation heating system bringing the load up to the final austenitizing temperature.

Gas Quenching

With gas quenching, the main priorities for all users are two-fold:

① to have the fastest quench rate possible;

② to have gas flow variability, allowing the adaption of flow patterns to the part and load geometry in order to achieve minimal distortion.

The reduction of distortion produced by gas quenching was largely advanced in the early 1980's, when single-chamber vacuum furnaces, combining high-pressure quenching systems with alternating gas flow direction, were developed. Thus, vertical gas flow through the load during high-pressure gas quenching, with upwards and downwards direction alternating in a time-or temperature-controlled manner, was the standard until 1996.

Certain loads with components forming a more shelf-type configuration, having inherently low penetrability for a vertical gas flow, required the development of an additional hori-zontal gas flow. Moreover, in the drive to increase the quality of the heat-treated components and expand the range of applications of single-chamber vacuum furnaces with high-pressure

gas quench systems, market demand called for gas-quench systems with improved performance.

This objective can be achieved in single-chamber vacuum furnaces by increasing the applied pressures, or making use of gases other than nitrogen; i. e. gases with higher heat conductivity and specific heat capacity, like helium or hydrogen. However, both factors, higher pressures and alternatives to nitrogen, increase the investment costs and the running costs of gas-quench systems considerably. Therefore, the need prevailed to engineer single-chamber vacuum furnaces with gas pressures limited to 10 bar maximum, but nevertheless possessing an improved quenching capability.

Carburizing in Vacuum Furnaces

Carburizing with hydrocarbon gases at sub-atmospheric pressures in vacuum furnaces is a process rapidly gaining acceptance, especially in conjunction with gas quenching. The outstanding features of this process are the rapid and high carbon transfer, the total lack of intergranular (internal) oxidation, and the singular bright surface quality associated with vacuum processes.

The high and rapid carbon transfer, totally beyond any equilibrium state, results in high surface carbon contents, frequently after only minutes of exposure to the hydrocarbon gas, and makes the control of the process difficult. The control methods used so far depend upon empirical measurements of carbon-transfer rates under various temperature and gas-pressure conditions. These rates are stored in data files and used in computational programs for carbon transfer and carbon diffusion. The control system normally stops the carburizing phase when the calculated surface carbon content reaches the saturation limit of the austenite. As this takes only minutes, standard low-pressure carburizing cycles are a sequence of multiple boost and diffuse stages up to the point where the given case depth is reached.

Two problems of the early low-pressure carburizing process of the 1970's were the lack of case depth uniformity and the production of soot in the surface. Both have been mainly overcome today by special gas injection techniques and reduced gas pressure (usually below 10 mbar).

Even in dense loads, case uniformities of 0. 1 mm for a case depth of 0. 6 mm are no longer a problem. If the component geometry becomes very complex, as in diesel injection systems, the assistance of plasma in an otherwise-unchanged low-pressure carburizing cycle promises to further enhance carbon penetration into small holes or passages. The evolution of plasma processes, like plasma carburizing and plasma nitriding, has been somewhat hampered in the past by the lack of plasma generators of sufficient current output, thus limiting the applicable load size. Today's availability of high-frequency pulsed generators, with current outputs of up to 1000A, has overcome this deficiency.

Outlook

The outlook for environmentally-friendly vacuum heat treatment processes and equipment seems to be very bright. Vacuum furnaces, with their close relationship to mechanical processing machines, are likely to be the solution for in-line integrated heat treatment.

Their increasing utilization will also stimulate wider use of elevated process temperatures, calling for the development of better temperature-resistant steel materials.

An absolute necessity for the wider use of low-pressure carburizing processes will be the development of in-situ monitoring of the carbon-transfer rate. It is predicted that sufficiently reliable and accurate sensors will be available within the next few years.

Gas quenching systems will also play a dominant role in future treatment lines, not only in conjunction with vacuum furnaces. Their economic success will be closely linked to the development of new constructional and casehardening steels with improved hardenability properties, making the gas-quench systems simpler and less costly.

Selected from "An Overview of Advances in Atmosphere and Vacuum Heat Treatment", B. Edenhofer, Heat Treatment of Metals, No. 1, 1999.

New Words and Expressions

1. carburization [ˌkɑːbjuraiˈzeiʃən] *n*. 渗碳，碳化
2. plasma [ˈplæzmə] *n*. [物] 等离子区，等离子体
3. convective [kənˈvektiv] *a*. 对流的，传递性的
4. harden [ˈhɑːdn] *vt*. 使硬化，使变硬，淬火
5. radiation [ˌreidiˈeiʃən] *n*. [物] 发光，发热，辐射，放射物，辐射能
6. austenitize [ˈɔːstənətaiz] *vt*. [冶] 奥氏体化，使产生奥氏体
7. two-fold 两倍，两方面
8. geometry [dʒiˈɔmitri] *n*. 几何学，几何图案
9. chamber [ˈtʃeimbə] *n*. 燃烧室，箱式，容器
10. intergranular [ˌintəˈgrænjulə] *a*. 晶粒间的
11. conjunction [kənˈdʒʌŋkʃən] *n*. 连接，结合
12. deficiency [diˈfiʃənsi] *n*. 缺乏，不足，缺陷，欠缺
13. hardenability [ˌhɑːdnəˈbiliti] *n*. 可硬性；[冶] 可淬性，淬透性

Unit 5 • Design of Machine and Machine Elements

Machine Design

Machine design is the art of planning or devising new or improved machines to accomplish specific purposes. In general, a machine will consist of a combination of several different mechanical elements properly designed and arranged to work together, as a whole. During the initial planning of a machine, fundamental decisions must be made concerning loading, type of kinematic elements to be used, and correct utilization of the properties of engineering materials. Economic considerations are usually of prime importance when the design of new machinery is undertaken. In general, the lowest over-all costs are designed. Consideration should be given not only to the cost of design, manufacture, sale and installation, but also to the cost of servicing. The machine should of course incorporate the necessary safety features and be of pleasing external appearance. The objective is to produce a machine which is not only sufficiently rugged to function properly for a reasonable life, but is at the same time cheap enough to be economically feasible①.

The engineer in charge of the design of a machine should not only have adequate technical training, but must be a man of sound judgment and wide experience, qualities which are usually acquired only after considerable time has been spent in actual professional work②.

Design of Machine Elements

The principles of design are, of course, universal. The same theory or equations may be applied to a very small part, as in an instrument, or, to a larger but similar part used in a piece of heavy equipment. In no case, however, should mathematical calculations be looked upon as absolute and final③. They are all subject to the accuracy of the various assumptions, which must necessarily be made in engineering work. Sometimes only a portion of the total number of parts in a machine are designed on the basis of analytic calculations. The form and size of the remaining parts are then usually determined by practical considerations. On the other hand, if the machine is very expensive, or if weight is a factor, as in airplanes, design computations may then be made for almost all the parts.

The purpose of the design calculations is, of course, to attempt to predict the stress or deformation in the part in order that it may safely carry the loads, which will be imposed on it, and that it may last for the expected life of the machine . All calculations are, of course, dependent on the physical properties of the construction materials as determined by laboratory tests. A rational method of design attempts to take the results of relatively simple and fundamental tests such as tension, compression, torsion, and fatigue and apply them to all the complicated and involved situations encountered in present-day machinery.

In addition, it has been amply proved that such details as surface condition, fillets, notches, manufacturing tolerances, and heat treatment have a market effect on the strength

and useful life of a machine part. The design and drafting departments must specify completely all such particulars, and thus exercise the necessary close control over the finished product.

As mentioned above, machine design is a vast field of engineering technology. As such, it begins with the conception of an idea and follows through the various phases of design analysis, manufacturing, marketing. The major areas of consideration in the general field of machine design are as follows:

1) Initial design conception
2) Strength analysis
3) Materials selection
4) Appearance
5) Manufacturability
6) Economy
7) Safety
8) Environment effects
9) Reliability and life
10) Legal considerations

Failure Analysis and Dimensional Determination

It is absolutely essential that a design engineer knows how and why parts fail so that reliable machines which require minimum maintenance can be designed. Sometimes, a failure can be serious, such as when a tire blows out on an automobile traveling at high speeds. On the other hand, a failure may be no more than a nuisance. An example is the loosening of the radiator hose in the automobile cooling system . The consequence of this latter failure is usually the loss of some radiator coolant, a condition which is readily detected and corrected.

The type of load a part absorbs is just as significant as the magnitude. Generally speaking, dynamic loads with direction reversals cause greater difficulties than static loads and, therefore, fatigue strength must be considered. Another concern is whether the material is ductile or brittle. For example, brittle materials are considered to be unacceptable where fatigue is involved.

Many people mistakingly interpret the word failure to mean the actual breakage of a part. However, a design engineer must consider a broader understanding of what constitutes failure. For example, a brittle material will fail under tensile load before any appreciable deformation occurs. A ductile material, however will deform a large amount prior to rupture. Excessive deformation, without fracture, may cause a machine to fail because the deformed part interferes with a moving second part. Therefore, a part fails (even if it has not physically broken) whenever it no longer fulfills its required function. Sometimes failure may be due to wear that can change the correct position of mating parts. The wear may be due to abnormal friction or vibration between two mating parts. Failure also may be due to a phenomenon called creep, which is the plastic flow of a material under load at elevated temperatures. In addition, the actual shape of a part may be responsible for failure. For example, stress concentrations due to sudden changes in contour must be taken into account. Evaluation of

stress considerations is especially important when there are dynamic loads with direction reversals and the material is not very ductile.

In general, the design engineer must consider all possible modes of failure, which include the following:

1) stress
2) deformation
3) wear
4) corrosion
5) vibration
6) environmental damage
7) loosening of fastening devices

The part sizes and shapes selected must also take into account many dimensional factors that produce external load effects, such as geometric discontinuities, residual stresses due to forming of desired contours, and the application of interference fit joints④.

Selected from "Design of Machine Elements (6th Edition)", M. F. Spotts, Prentice-Hall, Inc., 1985 and "Machine Design", Anthony Esposito, J. Robert Thrower, Delmar Publishers Inc., 1991.

New Words and Expressions

1. rugged [ˈrʌgid] *a.* 结实的，坚固的
2. fillet [ˈfilit] *n.* 嵌条，（内）圆角
3. notch [nɔtʃ] *n.*; *vt.* 槽口，凹口，刻痕；刻凹痕，开槽
4. tolerance [ˈtɔlərəns] *n.*; *vt.* 公差，容许量；给（机器部件等）规定公差
5. nuisance [ˈnjuːsns] *n.* 讨厌的人或东西，麻烦事，损害
6. loosen [ˈluːsn] *v.* 解开，放松，松开
7. radiator [ˈreidieitə] *n.* 散热器，电暖炉，辐射体
8. fracture [ˈfræktʃə] *v.*; *n.* （使）断裂，（使）破裂［碎］；断口［面］
9. wear [wɛə] *v.*; *n.* 磨损［蚀，破，坏］，损耗（量）
10. stress concentration 应力集中
11. geometric [dʒiəˈmetrik] *a.* 几何的，几何学的
12. discontinuity [ˈdisˌcɔntiˈnju(ː)iti] *n.* 间断，不连续，中断

Notes

①参考译文：目标是要生产一台机器，它不仅在合理寿命期内坚固耐用，同时价格低廉足以在经济上是可行的。

② 参考译文：负责设计一台机器的工程师，不仅应有足够的技术训练，而且还必须是一个有正确判断力和丰富经验的人，只有花费大量时间从事实际专业工作后才能获得这种素质。

③ 参考译文：然而，决不应该把数学计算看成是绝对的和决定性的。

④ 参考译文：所选零件大小和形状，也必须考虑到许多会产生外加载荷效应的尺寸因素，诸如几何不连续性、成形加工预期轮廓引起的残余应力以及过盈配合接头的应用。

Exercises

1. After reading the text above，write a summary of it.
2. Answer the following questions according to the text.
   ① What is the purpose of the design calculations?
   ② Please list the major areas of consideration in the general field of machine design.
   ③ Which possible modes of failure must the design engineer consider?
   ④ Explain the term "creep".
3. Translate the 1st paragraph into Chinese.
4. Put the following into Chinese by reference to the text.
   kinematic elements　external appearance　sound judgment　fatigue strength
   environmental damage　ductile or brittle　blow out　interference fit joints
5. Translate the following sentences into English.
   ① 设计计算的目的是预测零件的应力和变形，从而保证零件能安全承受施加的载荷，并保证在预期寿命内安全。
   ② 交变动载荷通常比静载荷更危险，必须考虑疲劳强度。

Reading Material 5

Design and Implementation of a Novel Dexterous Robotic Hand

Design Specifications

Finger Design，Number and Degrees of Freedom

As in many previous hand designs，the decision was made to use three fingers with the current design. This limits complexity and will produce secure，precise grasps of generic 3D objects if the fingertip design provides either sufficient friction forces or a form closure type of constraint. The hand will also be designed to allow alternate fingertips to be interchanged easily.

It is readily apparent that increasing the number of DOF of each finger will increase the dexterity and flexibility of the hand at the cost of increasing its complexity. Since a high degree of flexibility is the overall goal，each finger will have three computer controlled DOF (X, Y, Z). By actuating all three fingers in X-Y-Z，the hand will also allow the compliance of the held part to be actively controlled in 3D space (i. e. X, Y, Z, θ_x, θ_y and θ_z). As observed by many researchers，control of compliance is important for the successful automation of many manufacturing tasks，and of the insertion process in assembly in particular.

Scalability and Workspace

Although the hand will be designed to have a large workspace，realistically this workspace needs only to cover the range of part shapes and sizes encountered in the intended manufacturing task，i. e. the same hand would not be intended for use in electronics assembly and in aircraft assembly. At the same time it is intended that the design be scalable in terms of power and size to allow it to be used in many tasks.

The particularly challenging task for which the hand prototype will be built is automotive body-in-white assembly (BIWA). Since BIWA involves a large range of complex shaped parts and numerous dedicated tools, it is a good test case for the flexible hand design. Here the parts are typically less than 400 mm long and 75 mm deep, so a cylindrical workspace 400 mm dia. ×75 mm deep is ideal.

Force, Stiffness and Mass

The hand must produce sufficient internal grasping forces to secure the object without damaging it. For the BIWA application a force range of 10～150 N is expected to be adequate. In addition to resisting the grasping forces the hand's mechanical structure must withstand gravity, inertial and assembly forces. Taking the required accuracy of ±0.5 mm into account, and assuming the maximum externally applied force is ≤50 N, the required minimum stiffness is 100 N/mm.

A lower hand mass would allow a smaller, less expensive robot to be used. A tradeoff exists with this desirable characteristic and the stiffness, workspace and DOF requirements. Since the latter are greater contributors to the application flexibility of the hand, they will be given priority in the design process. The target for the hand's mass is 20 kg, which is within the typical payload capacity for a mid-size industrial robot.

Mechanical Design

The mechanical design of the finger mechanisms and hand frame focused on the need for both high stiffness and low mass. Due to space limitations, only the novel features of the designs will be highlighted. Its kinematic diagram is shown in Fig. 5.1, and overall configuration for the gripper is shown in Fig. 5.2.

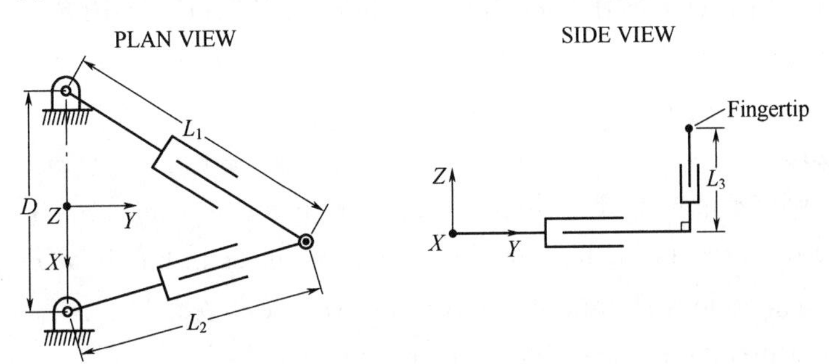

Fig. 5.1 Kinematic diagram for the hybrid parallel-serial finger mechanism

Finger Mechanism

The bipod parallel mechanism design provides high stiffness in the X-Y plane since its truss structure will place its members into compression and/or tension under load rather than bending. However, forces applied at the fingertip will create a bending moment which will bend the truss out of plane. To solve this problem a double truss structure was designed. With this structure, fingertip forces are resolved into reaction forces which are transmitted through each truss to the hand's frame.

To provide greater stiffness than possible with the actuators alone, brake mechanisms were added to allow the actuators to be locked. For compactness the brakes were incorporat-

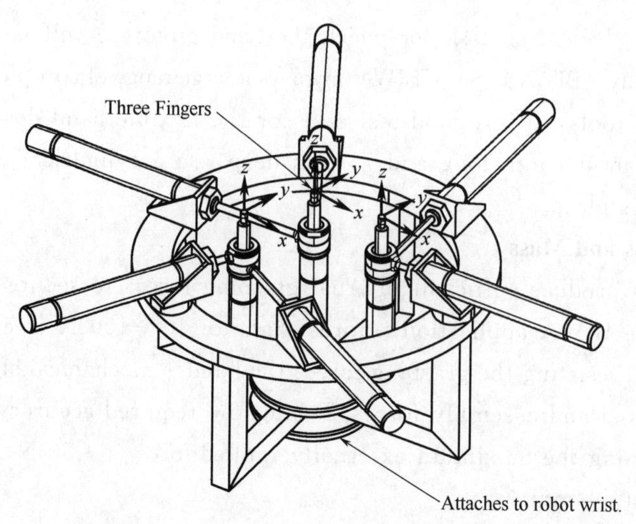

Fig. 5. 2 Overall configuration for the gripper

ed into the bearing blocks for the prismatic actuators. Four devices are used with each actuator. Each device consists of a 32 mm dia. pneumatic piston which clamps against one of the guide rods. Based on friction coefficient tests with a prototype, the total braking force per actuator is about 350 N.

For low mass and compactness, the design combines the middle revolute joint with the Z-axis prismatic actuator. The linear bearings and guide rod for the Z-axis are encircled by the ball bearings for the revolute joint . The Z-axis bearing block is rigidly mounted to one of the X-Y plane actuators, and a yoke is mounted to the second actuator. This yoke straddles the bearing block and preloads the revolute bearings to provide a stiff, slop free joint. The Z-axis guide rod forms the base of the hand finger. The fingertip is removable, allowing alternate designs to be interchanged as needed.

Frame Design

The frame will support the three finger mechanisms and attach to the Robot's wrist. With the requirements for high stiffness and low mass in mind, a welded space frame structure employing hollow aluminum extrusions was designed. The cross-sectional dimensions of the frame members necessary to meet the stiffness specification were determined from a theoretical analysis of the frame's stiffness.

Conclusion

A lightweight, large workspace, non-anthropomorphic dexterous hand was designed and implemented. To provide flexibility of application in manufacturing tasks, the design includes three fingers, each capable of being either position or force controlled in X-Y-Z. The conflicting workspace, strength, and mass requirements were met by designing a novel hybrid parallel-serial finger mechanism, a lightweight frame and pneumatic servos for finger actuation. The implemented design has a workspace of about 400 mm dia. × 75 mm deep, and a mass of 25 kg. A limitation of the design is that its overall size is roughly three times larger than its workspace so it is not well suited for grasping objects which are difficult to access.

Selected from "Design and Implementation of a Lightweight, Large Workspace, Non-Anthropomorphic Dexterous Hand", R. B. Van Varseveld, G. M. Bone, Journal of Mechanical Design, Vol. 121 (4), 1999.

New Words and Expressions

1. dexterous [ˈdekstərəs] *a*. 灵巧的，惯用右手的
2. generic [dʒiˈnerik] *a*. 属的，类的，一般的，普通的
3. fingertip [ˈfiŋgətip] *n*. 指尖，指套
4. dexterity [deksˈteriti] *n*. 灵巧，机敏
5. compliance [kəmˈplaiəns] *n*. 依从，顺从，顺应，柔顺
6. automation [ɔːtəˈmeiʃən] *n*. 自动化，自动操作
7. scalability [ˌskeiləˈbiliti] *n*. 可伸缩性，可缩放性，可量测性
8. prototype [ˈprəutətaip] *n*. 样机，原型
9. body-in-white assembly (BIWA) 白车身装配（是汽车行业中的特有术语，即未修饰喷漆前的汽车车身部件）
10. payload [ˈpeiləud] *n*. 净载质量，有效载荷
11. bipod [ˈbaipɔd] *n*. 两脚台，两脚架
12. bearing [ˈbεəriŋ] *n*. 轴承
13. prismatic [prizˈmætik] *a*. 棱形的，棱柱的，棱镜的
14. piston [ˈpistən] *n*. [机] 活塞，瓣
15. revolute [ˈrevəljuːt] *a*.; *vt*. 旋转的，向后卷的，外卷的；旋转
16. yoke [jəuk] *n*.; *v*. 轭，套，束缚；把……套上轭，连接
17. preload [priːləud] *vi*. 预加载，预装人
18. anthropomorphic [ænθrəpəuˈmɔːfik] *a*. 拟人的，有人形的，类人的
19. servo [ˈsəːvəu] *n*. 伺服，伺服系统，随动装置

Unit 6 • Manufacturing Process (1)

Computer-Aided Manufacturing

The scientific study of metal-cutting and automation techniques is products of the twentieth century. Two pioneers of these techniques were Frederick Taylor and Henry Ford. During the early 1900s, the improving U. S. standard of living brought a new high in personal wealth. The major result was the increased demand for durable goods. This increased demand meant that manufacturing could no longer be treated as a blacksmith trade, and the use of scientific study was employed in manufacturing analysis. Taylor pioneered studies in "scientific management" in which methods for production by both men and machines were studied. Taylor also conducted metal cutting experiments at the Midvale Steel Company that lasted 26 years and produced 400 tons of metal chips. The result of Taylor's metal-cutting experiments was the development of the Taylor tool-life equation that is still used in industry today. This tool-life equation is still the basis of determining economic metal cutting and has been used in adaptive controlled machining.

Henry Ford's contributions took a different turn from Taylor's. Ford refined and developed the use of assembly lines for the major component manufacturer of his automobile. Ford felt that every American family should have an automobile, and if they could be manufactured inexpensively enough then every family would buy one. Several mechanisms were developed at Ford to accommodate assembly lines. The automation that Ford developed was built into the hardware, and Ford realized that significant demand was necessary to offset the initial development and production costs of such systems.

Although manufacturing industries continued to evolve, it was not until the 1950s that the next major development occurred. For some time, strides to reduce human involvement in manufacturing were being taken. Specialty machines using cams and other "hardwired" logic controllers had been developed. The U. S. Air Force recognized the development time required to produce this special equipment and that the time required to make only small sequence changes was excessive. As a result, the Air Force commissioned the Massachusetts Institute of Technology to demonstrate programmable or numerically controlled (NC) machines (also known as "softwired" machines). With this first demonstration in 1952 came the beginning of a new era in manufacturing. Since then, digital computers have been used to produce input either in a directed manner to many NC machines, direct numerical control (DNC), or in a more dedicated control sense, computer numerical control (CNC). Today, machine control languages such as APT (Automatic Programming Tool) have become the standard for creating tool control for NC machines.

It is interesting to note that much of the evolution in manufacturing has come as a response to particular changes during different periods. For instance, the technology that evolves in the nineteenth century brought with it the need for higher-precision machining.

(This resulted in the creation of many new machine tools, a more refined machine design, and new production processes.) The early twentieth century became an era of prosperity and industrialization that created the demand necessary for mass-production techniques. In the 1950s it was estimated that as the speed of an aircraft increased, the cost of manufacturing the aircraft (because of geometric complexity) increased proportionately with the speed. The result of this was the development of NC technology.

A few tangential notes on this history include the following. As the volume of parts manufactured increases, the production cost for the parts decrease (this is generally known as "economy of scale"). Some of the change in production cost is due to fixed versus variable costs. For instance, if only a single part is to be produced (such as a space vehicle), all of the fixed costs for planning and design (both product and process) must be absorbed by the single item. If, however, several parts are produced, the fixed charge can be distributed over several parts. Changes in production cost, not reflected in this simple fixed-versus variable-cost relationship, are usually the result of different manufacturing procedures—transfer-line techniques for high-volume items versus job-shop procedures for low-volume items①.

Automated Manufacturing Systems

An automated manufacturing system consists of a collection of automatic or semiautomatic machines linked together by a "intrasystem" material-handling system. These systems have been around since before Henry Ford began to manufacture his Model T on his moving assembly line. These automated systems have been used to produce machined components, assemblies, electrical components, food products, chemical products, etc. The total number of products produced on a single system varies with the production methods. However, the principles of designing the production systems are the same independently of the product being manufactured. The workstations in a production system can be manual, semiautomatic or fully automatic. The automatic stations can be programmable or hardwired.

The purpose of any production system is to produce a product or family of products in the most economical manner. Automated production systems are no different from any other type of manufacturing system. In order to employ any form of automation, the implementation must be economically justifiable. Automation has traditionally been most appropriate for high-value products. However, flexible automation equipment has brought automation to some relatively low value products.

Selected from "Modern Manufacturing Process Engineering", Benjamin W. Niebel, *et. al.*, McGraw-Hill Publishing Company, 1989.

New Words and Expressions

1. Computer-Aided Manufacturing (CAM) 计算机辅助制造
2. blacksmith ['blæksmiθ] *n*. 锻工，铁匠
3. adaptive [ə'dæptiv] *a*. (自) 适应的
4. assembly [ə'sembli] *n*.; *v*. 装配
5. evolve [i'vɔlv] *v*. 进化，发展，进展
6. commission [kə'miʃən] *n*.; *v*. 代理，委托，委任

7. tangential [tænˈdʒenʃəl] *a*. 切线的，肤浅的，略为触及的
8. distribute [disˈtribjuːt] *v*. 分布
9. mass-production techniques 大批量生产技术
10. economy of scale 规模经济
11. versus [ˈvəːsəs] *prep*. 与……比较，……对……，作为……的函数
12. semiautomatic [ˈsemiɔːtəˈmætik] *a*. 半自动的
13. manual [ˈmænjuəl] *a*. 手工的

Notes

① 参考译文：并不反映在固定成本-可变成本这一简单关系中的生产成本变化，通常都是不同制造方法导致的结果——组合机床自动线技术用于大批量产品，机群制车间方式用于小批量产品。

Exercises

1. After reading the text above, summarize the main ideas of it in oral English.
2. Answer the following questions according to the text.
   ① What is Ford's main contribution?
   ② What causes the development of NC technology?
   ③ Why development of NC technology is regarded as beginning of a new era in manufacturing?
   ④ State the background of development of NC technology briefly.
3. Put the following expressions into Chinese by reference to the text.
   tangential notes　　flexible manufacturing system
   machine instruction　　economy of scale
   "hardwired" logic controller　　transfer-line
   numerically controlled (NC)　　direct numerical control (DNC)
   computer numerical control (CNC)
4. Translate the following into English.
   计算机辅助制造　　数控机床　　手工、半自动化或全自动化
   尽管机械制造业一直在持续的发展，但直到20世纪50年代才出现又一个重大进展。

Reading Material 6

Manufacturing Process (2)

What Is Process Planning?

Process planning has been defined as "the subsystem responsible for the conversion of design data to work instruction". A more specific definition of process planning is "that function within a manufacturing facility that establishes the processes and process parameters to be used (as well as those machines capable of performing these processes) in order to convert a piece-part from its initial form to a final form that is predetermined (usually by a design engineer) on a detailed engineering drawing." The input (raw) material to a process

may take a number of forms (in machining, these materials normally result from a metal-forming process; the most common of which are bar stock, castings, forgings, or perhaps just a slab of metal, other processes have other input materials). Another machining material might be a burn-out (a part produced by a flame-cutting operation) cut to some rough dimension, or just a rectangular block of material. This input material might have almost any shape and physical property. Some processes may alter the size or surface texture of a part. Other process, like heat-treating, change the physical properties of materials. More specifically, annealing would tend to lessen the material hardness and decrease the workpiece tensile strength.

With these types of raw materials as a base, the process planner must prepare a list of those (machining) processes needed to convert this normally predetermined material into its specified final geometry. The commonest metal-removal process that a process planner has at his disposal are turning, facing, milling, drilling, boring, broaching, shaping, gundrilling, reaming, planing, sawing, trepanning, burnishing, punching, and grinding. Some manufacturing people may consider some of the operations as subsets of a major category. Reaming is often considered a subset of drilling. Others may define further major categories. Some less-familiar process such as electric discharge machining (EDM), electrochemical machining (ECM), and laser machining are also used for material removal. All applicable processes that are available for production should be considered by the process planner.

Elements of Process Planning

Doyle separates the activities performed in process planning into seven general categories.

① Interpret the specification requirements.

② Position the part on the machine.

③ Determine the intermediate product requirements at each stage of processing.

④ Select the major pieces of equipment to handle the processing.

⑤ Select the tooling and sequence of processing steps within each operation.

⑥ Compute the process time requirements.

⑦ Document the process plan.

Some of these activities can be divided into smaller units; however, the activities described provided a convenient categorization for analysis.

Manual Process Planning

A process planner normally operates under the following constraints.

① He plans for a given set of machines.

② The machines are capable of a limited number of manufacturing operations.

③ The machines have a specific burden/workload.

Given these machine constraints, and a set of engineering drawings containing specific part—geometry requirements, the process planner relies on his experience to develop a set of process capable of producing a part. The selection of the process is neither entirely random nor totally predictable. There is usually more than one process capable of producing a specific surface: the process planner must choose what he believes is the best process. This is normally done by recalling similar parts, or at least similar surfaces, and the means utilized in manufacturing that part. In this manner, the planner compares specific process used to obtain some set of final specifications and chooses what he believes to be the best alternative. This type of planning is known as man-variant process planning and is the commonest type of planning used for production today.

Planning the operation to be used to produce a part requires knowledge of two groups of variables: the part requirements (as indicated by an engineering drawing); and the available machines and process, and the capabilities of each process. Given these variables, the planner selects the combination of processes required to produce a finished part. In selecting this combination of processes, a number of criteria are employed. Production cost or time are usually the dominant criteria in process selection; however, machine utilization and routing often affect the plans chosen. In general, the process planner tries to select the best set of processes and machines to produce a whole family of parts rather than just a single part.

As one might imagine, a large part of process planning, as it currently exists, is an art rather than a science. Process tolerance information is frequently recorded in a "black book". This information, may or may not be reliable. Similarly, the planner must also select operational information (such as machine speed, feed, and depth of cut). Again, a black book is employed to select these operating characteristics—this time the black book is somewhat more sophisticated (a machining data handbook); however, it does not guarantee operating efficiency, only feasibility.

Automated Process Planning

A practical process planning system with good decision rules for each activity will seldom generate bad plans. Such a practical system could be structured by reducing the decision-making process to a series of mechanical steps. However, even if a system of good decision rules were to be developed, the human interaction of man-variant process planning could still create problems. Process planning can become a boring and tedious job. Accordingly, man-variant planning often produces erroneous process plans. This, coupled with the labor intensity of man-variant planning, has led many industries to investigate the automation of process planning. Whether a planning system is to be automated or manual, the seven general requirements of process planning as defined by Doyle must be performed. In the following section, the framework for automated process planning will be described.

Spur and Optiz were among the first to write on the automation of manufacturing systems and the role that process planning should play in these systems. Spur was perhaps the first to define variant and generative methods of process planning and the mechanization and implementation of such planning systems. The variant method of process planning is based on the principle of group technology and essentially consists of two steps.

① Build a catalog (or "menu" as it is often called) of process plans to produce a gamut of parts, given a set of machine tools.

② Create the software necessary to examine the part that is being planned and find the closest facsimile in the catalogue, then retrieve the associated process plans.

The generative method of process planning essentially consists of four steps.

① Describe a part in detail.

② Describe a catalog of process available to produce parts.

③ Describe the machine tool(s) that can perform these processes.

④ Create the software to inspect the part, process, and available machinery to determine whether all three are compatible.

In general, planning using generative principles requires a detailed description of the part as well as a detailed understanding of manufacturing process and their accuracy. Manufacturing plans based on the variant principle are determined by activating several standard solutions for individual operations and adapting or adjusting them where necessary.

Selected from "Modern Manufacturing Process Engineering", Benjamin W. Niebel, *et. al.*, McGraw-Hill Publishing Company, 1989.

New Words and Expressions

1. metal-removal process 金属加工过程
2. turning ['tə:niŋ] *n.* 车削
3. facing ['feisiŋ] *n.* 车端面
4. milling ['miliŋ] *n.* 铣削
5. broach [brəutʃ] *v.* 拉削
6. trepanning [tri'pæniŋ] *n.* 穿孔
7. burnishing ['bə:niʃiŋ] *n.* 抛光
8. punching ['pʌntʃiŋ] *n.* 冲压
9. grinding ['graindiŋ] *n.* 磨削
10. feasibility [ˌfi:zə'biliti] *n.* 可行性
11. gamut ['gæmət] *n.* 范围
12. facsimile [fæk'simili] *n.* 复制品，传真
13. retrieve [ri'tri:v] *v.* 更正，恢复，取回，检索

Unit 7 • Shaper, Planer, Milling and Grinding Machines

Shapers

Shapers are machine tools used primarily in the production of flat and angular surfaces. In addition, the shaper is used to machine irregular shapes and contours which are difficult to produce on other machine tools. Internal as well as external surfaces and shapes can be produced on the shaper. Common shapes produced on the shaper are flat, angular, grooves, dovetails, T-slots, keyways, slots, serrations, and contours. Single-point cutting tools similar to the type used on the lathe are used in machining most surfaces on the shaper. Contour surfaces can be produced with either a single-point tool or a formed cutting tool. When using a single-point cutting tool for contour surfaces, the depth of cut must be constantly regulated by the machine operator to achieve the proper contour①. Profiling attachments can be used on some shapers to regulate the feed motion and provide duplication of other parts. Numerical control units also enable the production of irregular surfaces with constant regulation of the depth of cut.

Shapers are classified by the plane in which the cutting action occurs, either horizontal or vertical. In addition the horizontal-type shapers are further classified as push or pull cut. A push-cut shaper cuts while the ram is pushing the tool across the work, and a pull cut machine removes material while the tool is pulled toward the machine②. Vertical shapers use a pushing-type cutting action and are sometimes referred to as slotters or keyseaters.

Planers

Planers are similar to shapers because both machines are primarily used to produce flat and angular surfaces. However, planers are capable of accommodating much larger workpieces than the shaper. In planer operations the workpiece is mounted on the table which reciprocates in a horizontal plane providing a straight-line cutting and feed action. Single-point cutting tools are mounted on an overhead cross rail and along the vertically supported columns. The cutting tools are fed into or away from the workpiece on either the horizontal or vertical plane, thus being capable of four straight-line feed motions.

Cutting speeds are slow on the planer because of the workpiece size and type of cutting tool being used. In order to increase the production of the planer, multiple tooling stations are employed. Two tooling stations are located on the overhead cross rails, with usually one tooling station on the vertical supports. Another method of increasing production on the planer is to mount a number of workpieces on the table at the same time. This method is only feasible when the workpieces require the same cut and are relatively small in size. The planer size is designated by the maximum workpiece capacity of the machine. The height, width, and length of the workpiece that can be accommodated on the planer's worktable varies with the type of planer.

Milling Machines

Milling machines are probably the most versatile machine tools used in modern manufacturing with the exception of the lathe. Primarily designed to produce flat and angular surfaces, the milling machine is also used to machine irregular shapes, surfaces, grooves, and slots. The milling machine can also be used for drilling, boring, reaming, and gear-cutting operations.

A number of different types of milling machines are manufactured in order to serve the multitude of needs and the diverse applications. Milling machines are classified according to their structure and include column-knee, fixed bed, planer, and special machines.

Grinding Machines

Grinding machines utilize abrasive grains, bonded into various shapes and sizes of wheels and belts to be used as the cutting agent. Grinding operations are used to impart a high-quality surface finish on the workpiece. In addition, the dimensional accuracy of the workpiece is improved since tolerances of 0.00001 in. [0.00025mm] are possible in grinding operations. Both internal and external surfaces can be ground by using the variety of grinding machines available. Related operations—which use abrasives in various forms such as paste, powder, and grains—include lapping, honing, and drum finishing.

Grinding machines are classified according to the type of surface produced. Common surfaces and classifications of grinding machines are surface, cylindrical, and special machines.

Selected from "Basic Manufacturing Processes", H. C. Kazanas, Glenn E. Baker, *et al.*, McGraw-Hill Book Company, 1981.

New Words and Expressions

1. shaper ['ʃeipə] *n.* 牛头刨床，插床
2. planer ['pleinə] *n.* 龙门刨床
3. milling machine 铣床
4. grinding machine 磨床
5. contour ['kɔntuə] *n.* 轮廓，周线，等高线，外形，造型
6. dovetail ['dʌvteil] *n.* 楔形榫头
7. keyway ['kiːwei] *n.* 键（销）槽，销座，凹凸缝
8. serration [se'reiʃən] *n.* 齿面，锯齿（形，构造）
9. profiling ['prəufailiŋ] *n.* 仿形切削，靠模加工
10. duplication [ˌdjuːpli'keiʃən] *n.* 复制品，成倍
11. ram [ræm] *n.* （牛头刨）滑枕
12. keyseater ['kiːsiːtə] *n.* 键槽铣床，铣键槽机
13. reciprocate [ri'siprəkeit] *v.* （使）往复（运动）
14. versatile ['vəːsətail] *a.* 通用的，万向的，多能的
15. diverse [dai'vəːs] *a.* 不同的，（各种）各样的，多样的
16. abrasive [ə'breisiv] *n.* 研磨剂，磨料
17. bond [bɔnd] *v.* 黏结

18. impart [im'pɑːt] *v.* 给予，产生
19. lapping ['læpiŋ] *n.* 研磨，精磨
20. honing ['həuniŋ] *n.* 搪（珩）磨

Notes

① 参考译文：当对于周线表面使用一单点切削工具时，为取得恰当的周线，机器操作人员必须不断地调整切削深度。

the depth of cut must be constantly regulated by the machine operator 是该句的主句，to achieve the proper contour 是主句的行为目的，When using a single-point cutting tool for contour surfaces 是个“when ＋ing…”省略句形式，在 when 同 using 之间省略了 the machine operator is。

② 参考译文：当滑枕推着刀具横切过制品时，推式刨切削；而拉式刨刨去材料（是）在刀具被拉向刨床时。

句中 and 把两个都用 while 状语从句的句子连接为一个长句，and 把其前后句子转折性并列。句中 machine 意指 shaper，避免了用词重复。

Exercises

1. After reading the text above，summarize its main idea in oral English.
2. Answer the following questions according to the text.
   ① By what are shapers classified?
   ② Why are cutting speeds slow on the planer?
   ③ What is the milling machine used to do?
   ④ Where are abrasive bonded into?
3. Translate the 3rd and 7th paragraphs into Chinese.
4. Put the following into Chinese by reference to the text.
   workpiece　accommodate　impart　tolerance　paste
5. Put the following into English.
   规则形状　靠模［仿形］附件　高质量　卧式（机床）　垂直支撑
6. Translate the following sentences into English.
   ① 龙门刨床上增加产量的另一方法是在工作台上同时装上若干工件。
   ② 除了车床之外，铣床大概是当今制造业中被使用的最通用的机床。

Reading Material 7

Lathe

Machining Process Used to Produce Round Shapes

This chapter describes machining processes with the capability of producing parts that are basically round in shape. Typical products made are as small as miniature screws for the hinges of eyeglass frames and as large as turbine shafts for hydroelectric power plants，rolls

for rolling mills, and gun barrels.

One of the most basic machining processes is turning, meaning that the part is rotating while it is being machined. The starting material is usually a workpiece that has been made by other processes, such as casting, forging, extrusion, drawing, or powder metallurgy. Turning processes, which typically are carried out on a lathe or by similar machine tools, are outlined in Fig. 7. 1. These machines are highly versatile and capable of a number of machining processes that produce a wide variety of shapes as the following list indicates:

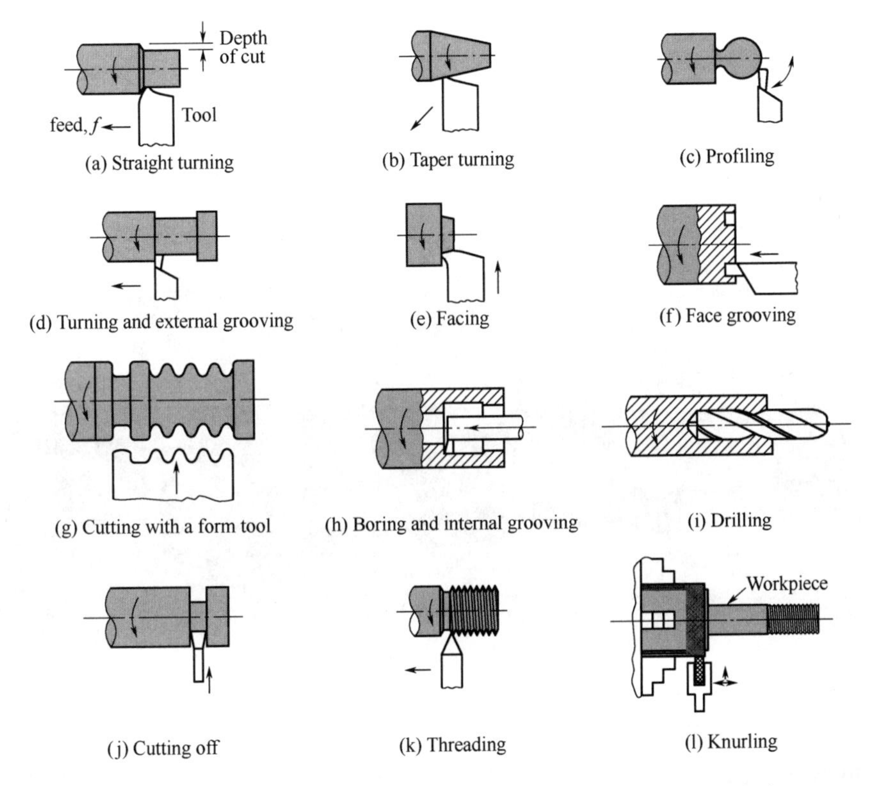

Fig. 7. 1 Various cutting operations that can be performed on a lathe

1. Turning, to produce straight, conical, curved, or grooved workpieces [Figs. 7. 1 (a)-(d)], such as shafts, spindles, and pins.

2. Facing, to produce a flat surface at the end of the part and perpendicular to its axis [Fig. 7. 1(e)], useful for parts that are assembled with other components. Face grooving produces grooves for applications such as O-ring seats [Fig. 7. 1(f)].

3. Cutting with form tools, [Fig. 7. 1(g)] to produce various axisymmetric shapes for functional or aesthetic purposes.

4. Boring, to enlarge a hole or cylindrical cavity made by a previous process or to produce circular internal grooves [Fig. 7. 1(h)].

5. Drilling, to produce a hole [Fig. 7. 1(i)], which may be followed by boring to improve its accuracy and surface finish.

6. Parting, also called cutting off, to cut a piece from the end of a part, as is done in the production of slugs or blanks for additional processing into discrete products [Fig. 7. 1(j)].

7. Threading, to produce external or internal threads [Fig. 7. 1(k)].

8. Knurling, to produce a regularly shaped roughness on cylindrical surfaces, as in making knobs [Fig. 7. 1(l)].

The cutting operations just summarized typically are performed on a lathe (Fig. 7. 2), which is available in a variety of designs, sizes, capacities, and computer-controlled features. Turning may be performed at various rotational speeds of the workpiece, depths of cut, and feeds, depending on the workpiece materials, cutting-tool materials, surface finish and dimensional accuracy required, and characteristics of the machine tool.

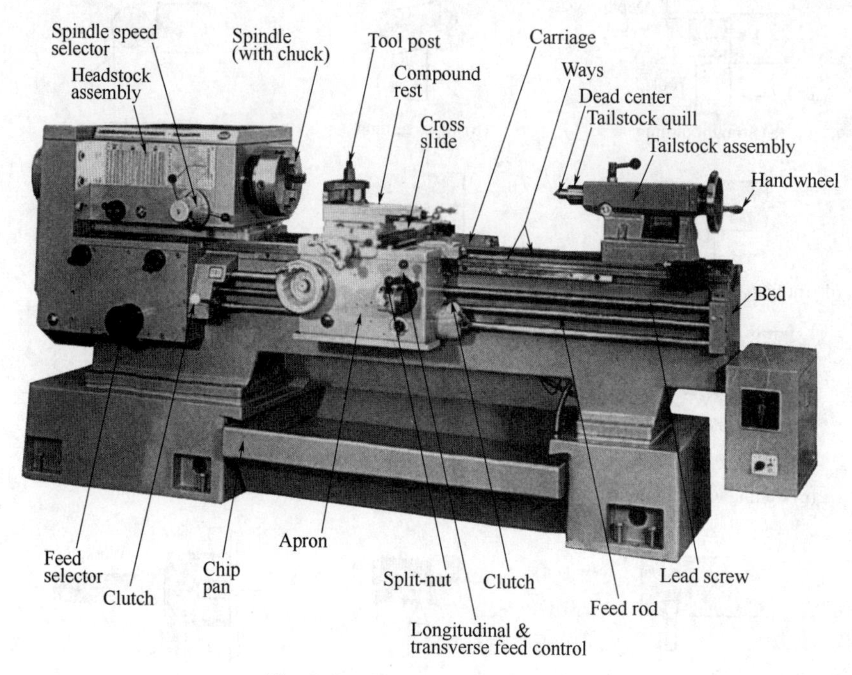

Fig. 7. 2 Components of a lathe

Lathe Components

Lathes are equipped with a variety of components and accessories, as shown in Fig. 7. 2. Their features and functions are as follows:

Bed. The bed supports all major components of the lathe. Beds have a large mass and are built rigidly, usually from gray or nodular cast iron. The top portion of the bed has two ways, with various cross-sections that are hardened and machined for wear resistance and dimensional accuracy during turning.

Carriage. The carriage, or carriage assembly, slides along the ways and consists of an assembly of the cross-slide, tool post, and apron. The cutting tool is mounted on the tool post, usually with a compound rest that swivels for tool positioning and adjustment. The cross-slide moves radially in and out, controlling the radial position of the cutting tool in operations such as facing [Fig. 7. 1(e)]. The apron is equipped with mechanisms for both manual and mechanized movement of the carriage and the cross-slide by means of the lead screw.

Headstock. The headstock is fixed to the bed and is equipped with motors, pulleys, and V-belts that supply power to the spindle at various rotational speeds. The speeds can be

set through manually controlled selectors or by electrical controls. Most headstocks are equipped with a set of gears, and some have various drives to provide a continuously variable speed range to the spindle. Headstocks have a hollow spindle to which workholding devices (such as chucks and collets) are mounted and long bars or tubing can be fed through them for various turning operations. The accuracy of the spindle is important for precision in turning, particularly in high-speed machining; preloaded tapered or ball bearings typically are used to rigidly support the spindle.

Tailstock. The tailstock, which can slide along the ways and be clamped at any position, supports the other end of the workpiece. It is equipped with a center that may be fixed (dead center) or may be free to rotate with the workpiece (live center). Drills and reamers can be mounted on the tailstock quill (a hollow cylindrical part with a tapered hole) to drill axial holes in the workpiece.

Feed Rod and Lead Screw. The feed rod is powered by a set of gears through the headstock. The rod rotates during the operation of the lathe and provides movement to the carriage and the cross-slide by means of gears, a friction clutch, and a keyway along the length of the rod. Closing a split nut around the lead screw engages it with the carriage; it is also used for cutting threads accurately.

Selected from "Manufacturing Engineering and Technology (Sixth edition in SI units)", Serope Kalpakjian *et al*, Prentice-Hall, Inc. 2009.

New Words and Expressions

1. parting ['pɑːtiŋ] *n*. 切断，分离
2. knurl [nəːl] *v*. ; *n*. 滚花，压花
3. carriage ['kræridʒ] *n*. （机床的）拖板，机器的滑动部分
4. cross-slide 横向滑板，横刀架，横拖板
5. tool post 刀架，刀座
6. apron ['eiprən] *n*. （机床刀座下的）溜板箱，拖板箱
7. compound rest 复式刀架，（车床）小刀架
8. lead screw 丝杆
9. headstock ['hedstɔk] *n*. 主轴箱，床头箱
10. collet ['kɔlit] *n*. 弹性夹头，套筒，套爪
11. tailstock ['teilstɔk] *n*. 尾架，尾座，顶尖座
12. quill [kwil] *n*. 钻杆，套管轴，空心轴，滚针，小镗杆
13. feed rod 进给杆，光杆

Unit 8 • Additive Manufacturing-3D Printing

3D printing is an additive manufacturing (AM) technique for fabricating a wide range of structures and complex geometries from three-dimensional (3D) model data. The process consists of printing successive layers of materials that are formed on top of each other. 3D printing, which involves various methods, materials and equipment, has evolved over the years and has the ability to transform manufacturing and logistics processes. Additive manufacturing has been widely applied in different industries, including construction, prototyping and biomechanical. The growing consensus of adapting the 3D manufacturing system over traditional techniques is attributed to several advantages including fabrication of complex geometry with high precision, maximum material savings, flexibility in design, and personal customization①.

This paper aims to provide a comprehensive review of 3D printing techniques in terms of the main methods employed.

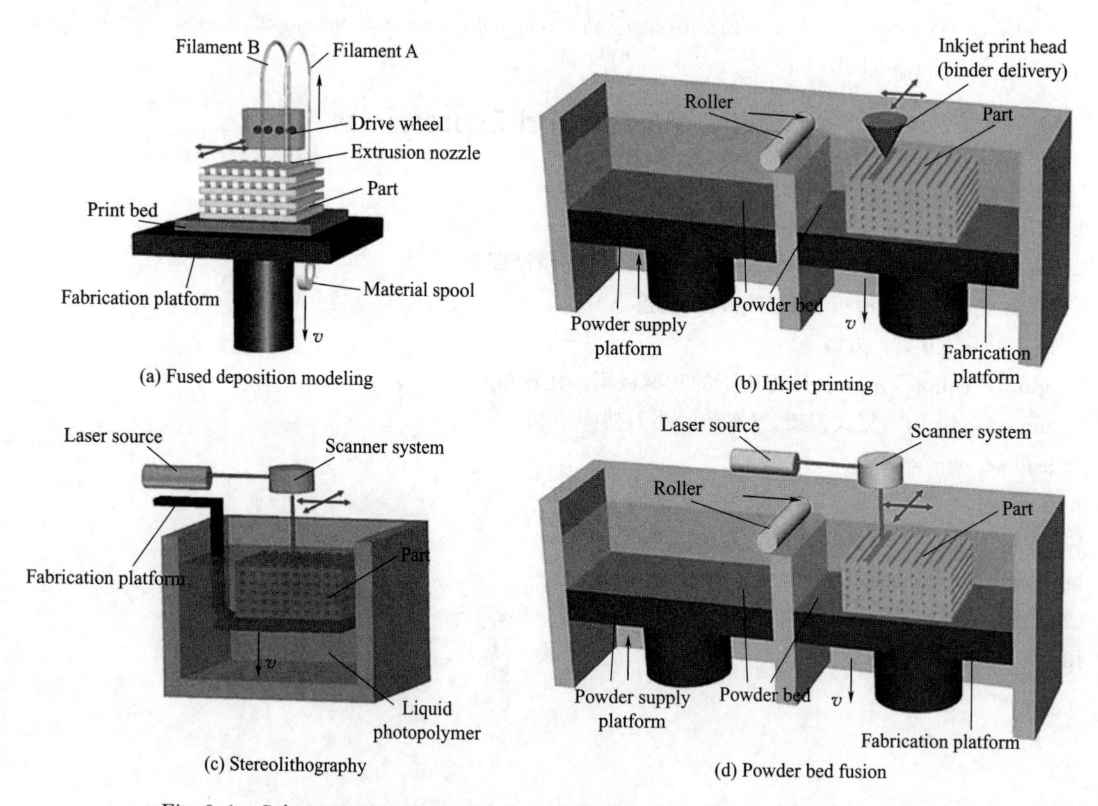

Fig. 8.1 Schematic diagrams of four main methods of additive manufacturing

Fused deposition modeling (FDM)

In FDM method, a continuous filament of a thermoplastic polymer is used to 3D print layers of materials [Fig. 8.1(a)]. The filament is heated at the nozzle to reach a semi-liquid

state and then extruded on the platform or on top of previously printed layers. The thermoplasticity of the polymer filament is an essential property for this method, which allows the filaments to fuse together during printing and then to solidify at room temperature after printing. The layer thickness, width and orientation of filaments and air gap (in the same layer or between layers) are the main processing parameters that affect the mechanical properties of printed parts. Inter-layer distortion was found to be the main cause of mechanical weakness. Low cost, high speed and simplicity of the process are the main benefits of FDM. On the other hand, weak mechanical properties, layer-by-layer appearance, poor surface quality and a limited number of thermoplastic materials are the main drawbacks of FDM. The development of fibre-reinforced composites using FDM has strengthened the mechanical properties of 3D printed parts. However, fibre orientation, bonding between the fibre and matrix and void formation are the main challenges that arise in 3D printed composite parts.

Inkjet printing and contour crafting

Inkjet printing is one of the main methods for the additive manufacturing of ceramics. In this method, a stable ceramic suspension e. g. zirconium oxide powder in water is pumped and deposited in the form of droplets via the injection nozzle onto the substrate. ② The droplets then form a continuous pattern which solidifies to sufficient strength in order to hold subsequent layers of printed materials③ [Fig. 8. 1(b)]. This method is fast and efficient, which adds flexibility for designing and printing complex structures. Two main types of ceramic inks are wax-based inks and liquid suspensions. The particle size distribution of ceramics, viscosity of the ink and solid content, as well as the extrusion rate, nozzle size and speed of printing, are factors that determine the quality of inkjet-printed parts. Maintaining workability, coarse resolution and lack of adhesion between layers are the main drawbacks of this method.

A similar technology to inkjet printing, called contour crafting, is the main method of additive manufacturing of large building structures. This method is capable of extruding concrete paste or soil by using larger nozzles and high pressure. Contour crafting has been prototyped to be used for construction on the moon.

Stereolithography (SLA)

SLA is one of the earliest methods of additive manufacturing, which was developed in 1986. It uses UV light (or electron beam) to initiate a chain reaction on a layer of resin or monomer solution. The monomers (mainly acrylic or epoxy-based) are UV-active and instantly convert to polymer chains after activation (radicalization). After polymerization, a pattern inside the resin layer is solidified in order to hold the subsequent layers [Fig. 8. 1(c)]. The unreacted resin is removed after the completion of printing. A post-process treatment such as heating or photo-curing may be used for some printed parts in order to achieve the desired mechanical performance. A dispersion of ceramic particles in monomers can be used to print ceramic-polymer composites or polymer-derived ceramifiable monomers e. g. silicon oxycarbide. SLA prints high-quality parts at a fine resolution as low as 10 μm. On the other hand, it is relatively slow, expensive and the range of materials for printing is very limited. Also,

the kinetics of the reaction and the curing process are complex. The energy of the light source and exposure are the main factors controlling the thickness of each layer. SLA can be effectively used for the additive manufacturing of complex nanocomposites.

Powder bed fusion

Powder bed fusion processes consist of thin layers of very fine powders, which are spread and closely packed on a platform. The powders in each layer are fused together with a laser beam or a binder. Subsequent layers of powders are rolled on top of previous layers and fused together until the final 3D part is built [Fig. 8. 1(d)]. The excess powder is then removed by a vacuum and if necessary, further processing and detailing such as coating, sintering or infiltration are carried out. Powder size distribution and packing, which determine the density of the printed part, are the most crucial factors to the efficacy of this method. Fine resolution and high quality of printing are the main advantages of powder bed fusion, which make it suitable for printing complex structures. This method is widely used in various industries for advanced applications such as scaffolds for tissue engineering, lattices, aerospace and electronics. The main advantage of this method is that the powder bed is used as the support, which overcomes difficulties in removing supporting material. However, the main drawbacks of powder bed fusion, which is a slow process, include high costs and high porosity when the powder is fused with a binder.

Direct energy deposition

Direct energy deposition (DED) has been used for manufacturing high-performance super-alloys. DED uses a source of energy (laser or electron beam) which is directly focused on a small region of the substrate and is also used to melt a feedstock material (powder or wire) simultaneously. The melted material is then deposited and fused into the melted substrate and solidified after movement of the laser beam. In this method, no powder bed is used and the feedstock is melted before deposition in a layer-by-layer fashion similar to FDM but with an extremely higher amount of energy for melting metals. Therefore, it can be helpful for filling cracks and retrofitting manufactured parts for which the application of the powder-bed method is limited. This method allows for both multiple-axis deposition and multiple materials at the same time. Moreover, DED can be combined easily with conventional subtractive processes to complete machining. This technique is commonly used with titanium, Inconel, stainless steel, aluminium and the related alloys for aerospace applications. DED can reduce the manufacturing time and cost, and provides excellent mechanical properties, controlled microstructure and accurate composition control. This method can be used for repairing turbine engines and other niche applications in various industries such automotive and aerospace.

Laminated object manufacturing

Laminated object manufacturing (LOM) is one of the first commercially available additive manufacturing methods, which is based on layer-by-layer cutting and lamination of sheets or rolls of materials. Successive layers are cut precisely using a mechanical cutter or laser and are then bonded together (form-then-bond) or vice versa (bond-then-form). The

form-then-bond method is particularly useful for thermal bonding of ceramics and metallic materials, which also facilitates the construction of internal features by removing excess materials before bonding. The excess materials after cutting are left for the support and after completion of the process, can be removed and recycled. LOM can be used for a variety of materials such as polymer composites, ceramics, paper and metal-filled tapes. Post-processing such as high-temperature treatment may be required depending on the type of materials and desired properties. Ultrasonic additive manufacturing (UAM) is a new subclass of LOM which combines ultrasonic metal seam welding and CNC milling in the lamination process. UAM is the only additive manufacturing method that is capable of construction of metal structures at low temperature. LOM has been used in various industries such as paper manufacturing, foundry industries, electronics and smart structures. Smart structures are classified as structures (which can be multi-tasking) with a number of sensors and processors. Unlike conventional methods, UAM can specify cavities in the structure based on the integrated computer design for embedded electronic devices, sensors, pipes and other features. Electronic devices can be printed in the same lamination process of UAM using direct write technologies. LOM can result in a reduction of tooling cost and manufacturing time, and is one of the best additive manufacturing methods for larger structures. However, LOM has inferior surface quality (without post-processing) and its dimensional accuracy is lower compared to the powder-bed methods. Also, removing the excess parts of laminates after formation of the object is time-consuming compared to the powder-bed methods. Therefore, it is not recommended for complex shapes.

Selected from "Additive manufacturing (3D printing): A review of materials, methods, applications and challenges", T. D. Ngo, *et al*. Composites Part B: Engineering. 2018, 143: 172-196.

New Words and Expressions

1. fused deposition modeling 熔融沉积造型
2. filament [ˈfiləmənt] *n*. 细丝，丝状物，单纤维
3. inkjet printing and contour crafting 喷墨打印与轮廓加工
4. zirconium [zɜːˈkəuniəm] *n*. 锆
5. substrate [ˈsʌbstreit] *n*. 基质，基材，底层，基底，底物
6. resolution [ˌrezəˈluːʃn] *n*. 分辨率，清晰度；决定，解决
7. stereolithography 激光立体印刷术，光固化立体造型
8. acrylic [əˈkrilik] *a*.; *n*. 丙烯酸的；丙烯酸
9. activation [ˌæktəˈveiʃən] *n*. 活化（作用），活性（化）
10. polymer-derived ceramifiable monomers 先驱体转化法制备陶瓷
11. silicon oxycarbide 碳氧化硅
12. nanocomposite [ˌnænə(ʊ)ˈkɔmpəzit] *n*. 纳米复合材料
13. powder bed fusion 粉末床融合
14. scaffold [ˈskæfəuld] *n*. 支架
15. lattice [ˈlætis] *n*. 格子架，斜条结构
16. sintering [ˈsintəriŋ] *n*. 烧结，熔结

17. infiltration [ˌinfilˈtreiʃən] *n.* 渗入，渗透，渗透物
18. direct energy deposition 直接能量沉积
19. niche [niːʃ] *n.* 适当的位置（场所）
20. laminated object manufacturing 叠层实体制造

Notes

① 参考译文：与传统技术相比，采用3D制造系统越来越成为共识，这归功于它的几个优势，包括：高精度制造复杂几何结构、最大限度地节省材料、设计的灵活性和个性化定制。

② 参考译文：在这个方法中，稳定的陶瓷悬浮液（如：氧化锆粉末水溶液）被泵送并通过注射喷嘴以液滴的形式沉积在基底上。

③ 参考译文：然后液滴形成连续的图案，固化至足够的强度得以支撑后续的打印材料层。

Exercises

1. After reading the text above，write an abstract of it.
2. Answer the following questions according to the text.
   ① What is the essential property for fused deposition modeling（FDM）?
   ② What are the main advantage and drawbacks of powder bed fusion ?
   ③ What is the basic principle of Stereolithography（SLA）?
   ④ List some examples of 3D printing applications.
3. Translate the 1st paragraph of the text into Chinese.
4. Put the following into Chinese by reference to the text.
   additive manufacturing　polymer filament　fusion　laser beam　solidify
   injection nozzle　contour crafting　embedded electronic device
5. Put the following into English.
   热塑性塑料　纤维增强复合材料　高分辨率　减材制造工艺　智能结构
   集成计算机　超声　光固化
6. Translate the following sentences into English.
   ① 在3D打印复合材料部件技术中，纤维取向、纤维与基体的结合以及孔隙的形成是主要的挑战。
   ② 先成型后黏合法特别适用于陶瓷和金属材料的热黏合，这也有助于通过在黏合前去除多余材料来确保内部性能。

Reading Material 8

Nontraditional Machining Processes

Conventional machining processes (i. e. , turning，drilling，milling) use a sharp cutting tool to form a chip from the work by shear deformation. In addition to these conventional methods，there is a group of processes that uses other mechanisms to remove material. The term nontraditional machining refers to this group that removes excess material by various techniques

involving mechanical, thermal, electrical, or chemical energy (or combinations of these energies).

The nontraditional processes have been developed since World War Ⅱ largely in response to new and unusual machining requirements that could not be satisfied by conventional methods. These requirements, and the resulting commercial and technological importance of the non-traditional processes, include:

- The need to machine newly developed metals and nonmetals. These new materials often have special properties (e. g. , high strength, high hardness, high toughness) that make them difficult or impossible to machine by conventional methods.
- The need for unusual and/or complex part geometries that cannot easily be accomplished and in some cases are impossible to achieve by conventional machining.
- The need to avoid surface damage that often accompanies the stresses created by conventional machining.

There are literally dozens of nontraditional machining processes, most of which are unique in their range of applications. In the present chapter, we discuss those that are most important commercially.

Ultrasonic machining (USM)

Ultrasonic machining (USM) is a nontraditional machining process in which abrasives contained in a slurry are driven at high velocity against the work by a tool vibrating at low amplitude and high frequency. The amplitudes are around 0. 075 mm (0. 003 in), and the frequencies are approximately 20,000 Hz. The tool oscillates in a direction perpendicular to the work surface, and is fed slowly into the work, so that the shape of the tool is formed in the part. However, it is the action of the abrasives, impinging against the work surface, that performs the cutting. The general arrangement of the USM process is depicted in Fig. 8. 2.

Common tool materials used in USM include soft steel and stainless steel. Abrasive materials in USM include boron nitride, boron carbide, aluminum oxide, silicon carbide, and diamond. Grit size ranges between 100 and 2,000. The vibration amplitude should be set approximately equal to the grit size, and the gap size should be maintained at about two times grit size. To a significant degree, grit size determines the surface finish on the new work surface. In addition to surface finish, material removal rate is an important performance variable in ultrasonic machining. For a given work material, the removal rate in USM increases with increasing frequency and amplitude of vibration.

Electrochemical machining (ECM)

In effect, electrochemical machining processes are the reverse of electroplating. The work material must be a conductor in the electrochemical machining processes. Electrochemical machining removes metal from an electrically conductive workpiece by anodic dissolution, in which the shape of the workpiece is obtained by a formed electrode tool in close proximity to, but separated from, the work by a rapidly flowing electrolyte. ECM is basically a deplating operation. As illustrated in Fig. 8. 3, the workpiece is the anode, and the tool is the cath-

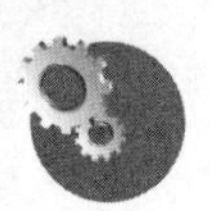

ode. The principle underlying the process is that material is deplated from the anode (the positive pole) and deposited onto the cathode (the negative pole) in the presence of an electrolyte bath. The difference in ECM is that the electrolyte bath flows rapidly between the two poles to carry off the deplated material, so that it does not become plated onto the tool.

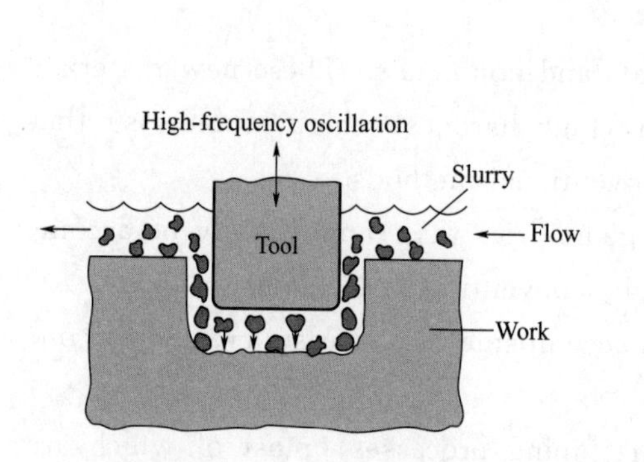

Fig. 8. 2 Ultrasonic machining

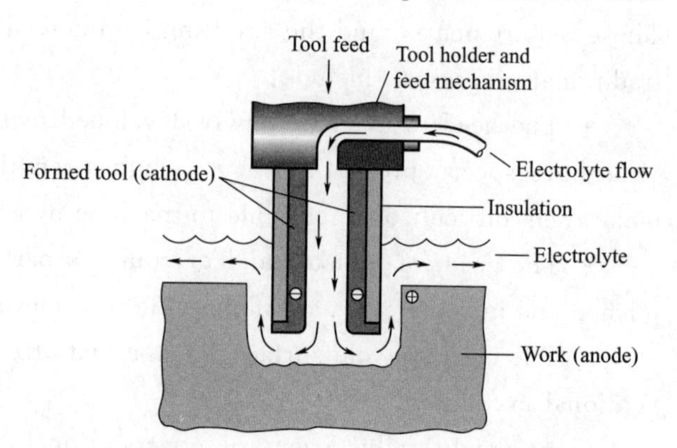

Fig. 8. 3 Electrochemical machining

Electric discharge machining (EDM)

Electric discharge machining (EDM) is one of the most widely used nontraditional processes. These processes remove metal by a series of discrete electrical discharges (sparks) that cause localized temperatures high enough to melt or vaporize the metal in the immediate vicinity of the discharge. An EDM setup is illustrated in Fig. 8. 4. The shape or the finished work surface is produced by a formed electrode tool. The sparks occur across a small gap between tool and work surface. The discharges are generated by a pulsating direct current power supply connected to the work and the tool.

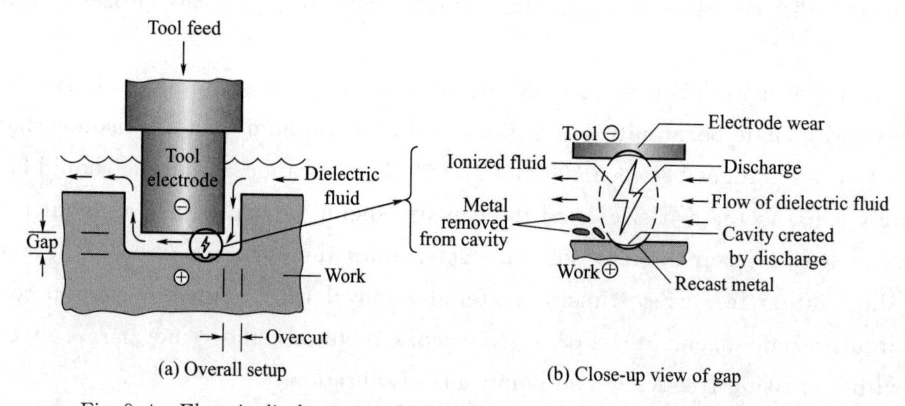

Fig. 8. 4 Electric discharge machining showing discharge and metal removal

Fig. 8. 4(b) shows a close-up view of the gap between the tool and the work. The discharge occurs at the location where the two surfaces are closest. The dielectric fluid ionizes at this location to create a path for the discharge. The region in which discharge occurs is heated to extremely high temperatures, so that a small portion of the work surface is suddenly melted and removed. The flowing dielectric then flushes away the small particle (call it a "chip"). Because the surface of the work at the location of the previous discharge is now

separated from the tool by a greater distance, this location is less likely to be the site of another spark until the surrounding regions have been reduced to the same level or below. Although the individual discharges remove metal at very localized points, they occur hundreds or thousands of times per second so that a gradual erosion of the entire surface occurs in the area of the gap.

Laser beam machining (LBM)

Lasers are being used for a variety of industrial applications. The term *laser* stands for light amplification by stimulated emission of radiation. A laser is an optical transducer that converts electrical energy into a highly coherent light beam. Laser beam machining (LBM) uses the light energy from a laser to remove material by vaporization and ablation. The setup for LBM is illustrated in Fig. 8. 5. The light beam is pulsed so that the released energy results in an impulse against the work surface that produces a combination of evaporation and melting, with the melted material evacuating the surface at high velocity.

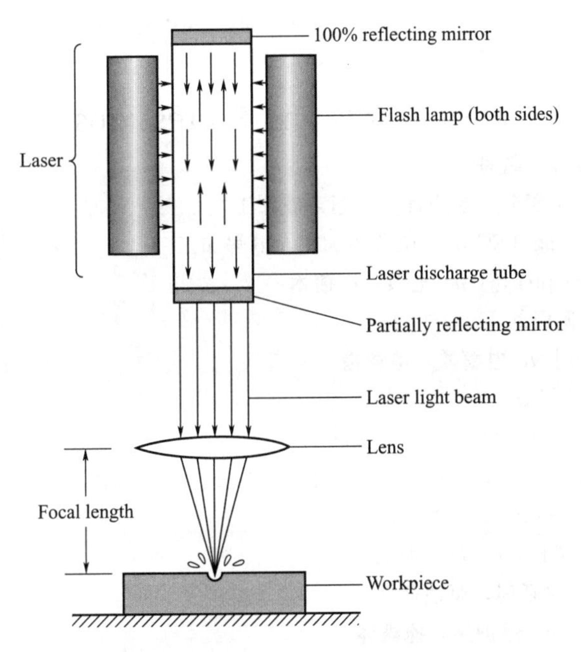

Fig. 8. 5 Laser beam machining (LBM)

LBM is used to perform various types of drilling, slitting, slotting, scribing, and marking operations. LBM is not considered a mass production process, and it is generally used on thin stock. The range of work materials that can be machined by LBM is virtually unlimited. Ideal properties of a material for LBM include high light energy absorption, poor reflectivity, good thermal conductivity, low specific heat, low heat of fusion, and low heat of vaporization. Of course no material has this ideal combination of properties. The actual list of work materials processed by LBM includes metal with high hardness and strength, soft metals, ceramics, glass and glass epoxy, plastics, rubber, cloth, and wood.

Chemical machining (CHM)

Chemical machining (CHM) is a nontraditional process in which material is removed by

means of a strong chemical etchant. The chemical machining process consists of several steps. They are:

① Cleaning. The first step is a cleaning operation to ensure that material will be removed uniformly from the surfaces to be etched.

② Masking. A protective coating called a maskant is applied to certain portions of the part surface. This maskant is made of a material that is chemically resistant to the etchant. It is therefore applied to those portions of the work surface that are not to be etched.

③ Etching. This is the material removal step. The part is immersed in an etchant that chemically attacks those portions of the part surface that are not masked. The usual method of attack is to convert the work material (e.g.. a metal) into a salt that dissolves in the etchant and is thereby removed from the surface. When the desired amount of material has been removed, the part is withdrawn from the etchant and washed to stop the process.

④ Demasking. The maskant is removed from the part.

Selected from "Fundamentals of Modern Manufacturing: Materials, Processes, and Systems (Fifth Edition)", Mikell P. Groover, John Wiley & Sons, Inc. 2013.

New Words and Expressions

1. nontraditional machining 特种加工
2. ultrasonic machining (USM) 超声加工，超声波加工
3. electrochemical machining (ECM) 电化学加工，电解加工
4. electroplating [iˈlektrəupleitiŋ] *n.* 电镀，电镀术
5. anodic [æˈnɔdik] *a.* 阳极的
6. electrolyte [iˈlektrəlait] *n.* 电解质，电解液
7. cathode [ˈkæθəud] *n.* 阴极
8. electric discharge machining (EDM) 电火花加工
9. dielectric [daiiˈlektrik] *a.*; *n.* 绝缘的，不导电的；电介质；绝缘体
10. laser beam machining (LBM) 激光束加工
11. chemical machining (CHM) 化学加工
12. etchant [ˈetʃənt] *n.* 蚀刻剂，腐蚀剂
13. maskant [ˈmɑskænt] *n.* 保护层，掩蔽体

Unit 9 • NC Machines

Machine may be classified according to the number of axes of numerically controlled movements with respect to Cartesian *X-Y-Z* coordinates. There may be other movement not numerically controlled. A two-axis machine would have the table moved lengthwise and crosswise in a horizontal plane; a three-axis machine would have an additional vertical movement of the spindle, for example, Four-, five-, or six-axis machines provide additional linear or rotary movements.

Three classes of NC systems are commonly recognized. One of these is called point-to-point or positioning NC, where a cutting tool and workpiece are positioned with respect to each other before a cut is taken, as exemplified by an NC drill press①. The path between points is of little concern and is not particularly controlled. Another class, known as a straight-cut NC system, involves movements between points like point-to-point NC but along straight or curved paths determined by the machine ways or slides. An NC turret lathe may be of the second class. A third class is continuous-path NC, or more formally continuous tool path control, which does contouring or profiling of lines, curves, or surfaces, of all shapes. One way of achieving continuous path NC is to move the workpiece or cutter from point to point along a straight line or curve (a parabolic path between points is used on some machines) with many points spaced closely enough together so that the composition path of the cut approximates a desired curve within limits required②. Another way is to drive machine members, say a saddle and the table on it, along coordinate axes at varying velocities controlled so that the resultant motion is along a particular line or curve.

An NC machine may be controlled through an open-loop or closed-loop circuit as depicted in Fig. 9. 1. The open-loop system is the simplest and cheapest but does not assure accuracy. A signal that is an order for a certain action (such as to turn on the coolant, start the spindle, or move the table to a certain position) is issued by the control unit. This travels through the drive mechanism which imposes the action upon the controlled member of the machine. The basic drive mechanism for a table or other machine member in an open-loop system is a stepping motor, also called a digital or pulse motor. It has, for example, 49 poles around the armature inside of 50 poles around the stator. Two poles are always aligned. Poles are energized to align the next two poles and turn the armature through a definite small angle each time a pulse of electricity is received. The motor turns the leadscrew, which moves the table a corresponding distance. If, for example, a pulse stands for a 0.001 mm movement, and a movement of 0.05 mm is desired, the control unit issues 50 pulses; the motor turns through 50 steps and turns the leadscrew, and the table (or other member) is driven the 0.05 mm required. The rate at which the pulses are issued determines the rate of feed. The power of a stepping motor is limited, and if the resistance to movement is large, the motor may just stall and miss steps. There is no feedback to report the omission, and ac-

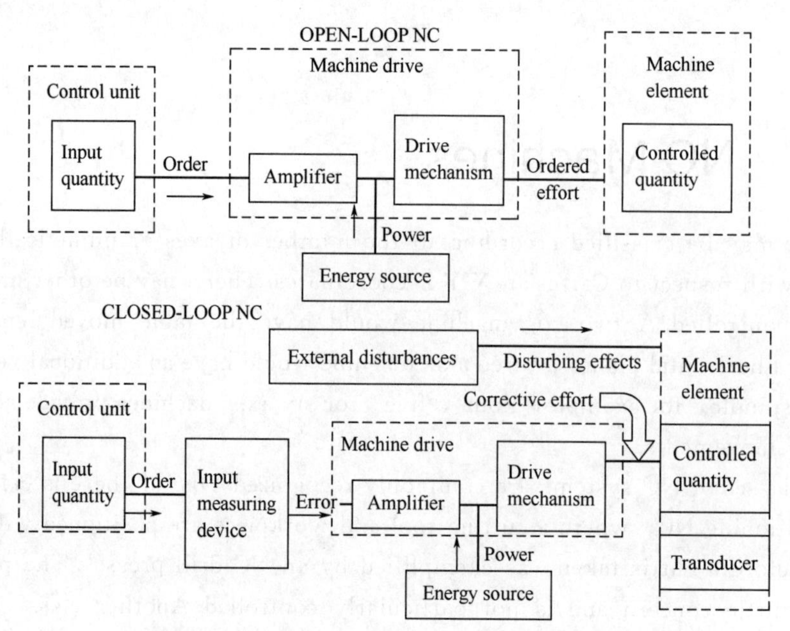

Fig 9. 1 Schematics of open-and closed- loop NC systems

curacy is lost. Direct drive with a stepping motor is confined to light service at fractional horsepower. For more power and heavy service, a stepping motor is matched with a hydraulic motor through a servo valve for table drives up to 7. 5 kW (10 hp) and more.

It is common in closed-loop systems for the pulsed command signal for the movement of the table (or other member) of the machine to be converted to a steady analog signal. That signal turns on the power to the drive mechanism. Hydraulic and ac but mostly dc electric motors are all used for NC machine drives. The intensity of the signal is determined by the feed velocity ordered. For a higher feed, pulses are issued at a faster rate and/or the analog signal is larger, which turns on more power to the motor to drive the table faster. A tachometer in the motor drive sends back a signal that is compared to the feed rate order to assure that the required feed rate is being delivered and to make corrections if it is not. Commonly, a resolver is driven by the leadscrew and returns a signal showing how far the table moves. More accuracy for a longer period of time is obtained with the feedback device attached directly to the table, but such devices are more costly. The feedback signal is compared to the order, and where the feedback equals the order and signifies the required movement has been completed, the power is shut off, and the movement stops. Since they are controlled through separate channels, all movements may act at the same time. Indeed, it is necessary that they act simultaneously and be synchronized accurately for contouring.

Selected from "Manufacturing Processes and Materials for Engineers" (THIRD EDITION), Lawrence E. Doyle, *et al.*, Prentice-Hall, Inc. 1985.

New Words and Expressions

1. **point-to-point** *a.* 点至点的，点位控制，定向的
2. **positioning** [pəziʃəniŋ] *n.* 定位，固定位置

3. straight-cut *n*. 纵向切削
4. continuous-path *n*. 连续路径
5. open-loop *n*. 开口［非闭合］回路，开环
6. closed-loop *n*. 闭合回路［电路，环路］，闭环
7. lengthwise [ˈleŋθˌwaiz] *a*. 纵向的
8. exemplify [igˈzemplifai] *vt*. 例证；作为……的例子
9. parabolic [ˌpærəˈbɔlik] *a*. 抛物线的，抛物面的
10. saddle [ˈsædl] *n*. 鞍状结构，滑动座架，滑座，床鞍
11. feedback [ˈfiːdbæk] *n*. 反馈，回复
12. fractional [ˈfrækʃənl] *n*. 分数的，分步的
13. servo valve 伺服阀，继动阀
14. analog [ˈænəlɔg] *n*. 模拟（量），比拟
15. AC＝alternating current *n*. 交流电
16. DC＝direct current *n*. 直流电
17. tachometer [tæˈkɔmitə] *n*. 转速计，流速计
18. armature [ˈɑːmətjuə] *n*. 电枢，（电机）转子，衔铁
19. synchronize [ˈsiŋkrənaiz] *v*. 同步化（使）同步，（使）同时（发生）

Notes

① 参考译文：（在三类数控系统中），其中一类被称为逐点或位置调节数控，在这类数控系统中，进行切削之前，切削刀具和工件彼此之间的位置就被规定了，数控钻床就是一个例子。

该句 where 之前为主句，where 引导一非限定性定语从句。

② 参考译文：得到数控连续路径的一个方式是使工件或刀具沿一直线或曲线作点到点的运动（在某些机床上，点之间采用抛物线路径），许多点的间隔足够小，以便在要求的限度内使切削组合的路径接近一预想的曲线。

这个句子以 so that 引导目的状语从句，so that 前为该句的主句，且以动词不定式 to move 为主句的表语。

Exercises

1. After reading the text above，summarize its main idea in oral English.
2. Answer the following questions according to the text.
   ① What is an NC machine controlled through?
   ② How many poles are needed around the armature?
   ③ Why is accuracy lost if the resistance to movement is large，and a stepping motor is used?
   ④ What signal does a resolver return when it is driven by the leadscrew?
3. Translate the 1st and 4th paragraphs into Chinese.
4. Put the following into Chinese by reference to the text.

Cartesian coordinates	digital or pulse motor	straight-cut NC system
with respect to each other	from point to point	

5. Put the following into English.

控制单元　　数控系统　　闭路系统　　电机　　解数器

6. Translate the following sentences into English.

① 在开口回路系统中，一工作台或其他机器构件的基本驱动装置是步进电机，也叫数值或脉动电机。

② 发出脉冲的频率确定（走刀量）给进的频率。

Reading Material 9

Numerical Control

Numerical control refers to the operation of machine tools from numerical data stored on paper or magnetic tape, tabulating cards, computer storage disks, or direct computer information. A historical example of using instructions punched on paper tape is the player piano. Notes to be played (instructions) are defined as a series of holes on a piano roll (punched paper tape), then sensed by the piano (using a pneumatic system powered by a foot-operated bellows), which plays the notes (executes the instructions).

Because mathematical information is used, the concept is called numerical control (NC), NC is the operation of machine tools and other processing machines by a series of coded instructions. Perhaps the most important instruction is the relative positioning of the tool to the workpiece. An organized list of commands constitutes an NC program. The program may be used repeatedly to obtain identical results. Manual operation of machine tools may be unsurpassable in producing fine-quality work, but such qualities are not consistent. NC is not a machine method; it is a means for machine control. NC is considered as one of the most dramatic and productive developments in manufacturing in this century.

Numerical Control Machine Tools

Early NC design placed control units on existing machine tool structures to accomplish numerical control. As experience was gained, it was apparent that NC machines were more efficient in the overall operation than conventional machines. But the controls became more adaptable to production equipment, and most of the equipment shown in this book has the potential for numerical control. Inspection, pipe bending, flame-cutting, wire wrapping, circuit board stuffing of electronic chips, laser cutting of fabric, drafting machines, and production processes have proven applications. In some instances the added controls cost more than the basic machine tool, but it is often a worthwhile expense. Solid-state circuitry has provided more reliable control at lower cost than previous electronic technology.

NC should be considered whenever there is similar raw material and work parts are produced in various sizes and complex geometries. Applications are in low-to medium-quantity batches, and similar processing sequences are required on each workpiece. Those production shops having frequent changeovers will benefit.

NC machine tools incorporate many advances such as programmed optimization of

cutting speeds and feeds, work positioning, tool selection, and chip disposal. The adoption of NC has altered existing designs to the point that NC machine tools have their own characteristics separate from the machine tools described in the various chapters. For example, modifications to the turret lathe have resulted in a turret slanted on the back side rather than placed on the front horizontal ways. A greater number of tools can be mounted on the turret as a result of the structural adaptation.

Development of the machining center with tool storage resulted from NC. Each tool can be selected and used as programmed. These machining centers can do almost all types of machining such as milling, drilling, boring, facing, spotting, and counterboring. Some machining operations can be programmed to occur simultaneously. The NC program selects and returns cutting tools to and from the storage magazine, if equipped, and also inserts them into a spindle. Parts can be loaded and moved between pallets, manipulated by rotation, and inspected after the work is finished. Robotic operation is possible, also being accomplished by NC.

Operational Sequence

NC starts with the parts programmer who, after studying the engineering drawing, visualizes the operations needed to machine the workpiece. These instructions, commonly called a program, are prepared before the part is manufactured and consist of a sequence of symbolic codes that specify the desired action of the tool workpiece and machine. Even in computer aided design and manufacturing this interpretation is necessary. The engineering drawing of the workpiece is examined, and processes are selected to transform the raw material into a finished part that meets the dimensions, tolerances, and specifications. This process planning is concerned with the preparation of an operations sheet or a route sheet or traveler. These different titles describe the procedure or the sequence of the operations, and it lists the machines, tools, and operational costs. The particular order is important. Once the operations are known, those that pertain to NC are further engineered in that detail sequences are selected.

A program is prepared by listing codes that define the sequence. A part programmer is trained about manufacturing processes and is knowledgeable of the steps required to machine a part, and documents these steps in a special format. These are two ways to program for NC, by manual or computer assisted part programming. The part programmer must understand the processor language used by the computer and the NC machine.

If manual programming is required, the machining instructions are listed on a form called a part program manuscript. This manuscript gives instructions for the cutter and workpiece, which must be positioned relative to each other for the path instructions to machine the design. Computer assisted part programming, on the other hand, does much of the calculation and translates brief instructions into a detailed instruction and coded language for the control tape. Complex geometries, many common hole centers, and symmetry of surface treatment can be simply programmed under computer assistance, which saves pro-

grammer time.

Tape preparation is next, as the program is "typed" onto a tape or punched card. If the programming is manual, the 1 in. (25 mm) wide perforated tape is prepared from the part manuscript on a typewriter with a standard keyboard but equipped with a punch device capable of punching holes along the length of the tape. If the computer is used, the internal memory interprets the programming steps, does the calculations to provide a listing of the NC steps, and additionally will prepare the tape. Some tapes contain electronic or magnetic signals; other systems use disks or direct computer inputs.

Verification is the next step, as the tape is run through a computer, and a plotter will simulate the movements of the tool and graphically display the final paper part often in a two-dimensional layout describing the final part dimensions. This verification uncovers major mistakes.

The final step is production using the NC tape. This involves ordering special tooling, fixtures, and scheduling the job. A machine operator loads the tape onto a program reader that is part of the machine control unit, often called a MCU. This converts coded instructions into machine tool actions. The media that the MCU can sense may be perforated tape, magnetic tape, tabulating cards, floppy disks, or direct computer signals from other computers or satellites. Perforated paper tape is the predominant input medium, but the concepts are the same whatever the input.

Selected from "Manufacturing Processes" (Eighth edition), B. H. Amstead, *et al.*, JOHN WILEY & SONS, Inc. 1987.

New Words and Expressions

1. **tabulating** ['tæbjuleitiŋ] *n.* 用表格表示
2. **pneumatic** [njuː'mætik] *a.* 气动的，空气的
3. **unsurpassable** [ˌʌnsə'pɑːsəbl] *a.* 无法超越的
4. **wrapping** ['ræpiŋ] *n.* 包（裹，装），打包
5. **stuffing** ['stʌfiŋ] *n.* 填料，填充剂，加脂
6. **batch** [bætʃ] *n.* 一次操作所需原料量，一次生产量，一批，批量
7. **disposal** [dis'pəuzəl] *n.* 处理，清除，处理方法
8. **slant** [slɑːnt] *v.*; *n.* 倾斜
9. **drilling** ['driliŋ] *n.* 钻孔，钻削
10. **boring** ['bɔːriŋ] *n.* 镗削，镗孔，扩孔
11. **spotting** ['spɔtiŋ] *n.* 钻中心孔，找正
12. **counterboring** ['kauntə'bɔːriŋ] *n.* 镗阶梯孔，镗孔，锪平底孔
13. **pallet** ['pælit] *n.* 板台，滑板，托板，托盘
14. **pertain** [pə'tein] *vi.* 从属于，适合

Unit 10 ● Hydraulic System

The history of hydraulic power is a long one, dating from man's prehistoric efforts to harness the energy in the world around him[①]. The only sources readily available were the water and the wind — two free and moving streams.

The watermill, the first hydraulic motor, was an early invention. One is pictured on a mosaic at the Great Palace in Byzantium, dating from the early fifth century. The mill had been built by the Romans. But the first record of a watermill goes back even further, to around 100 BC, and the origins may indeed have been much earlier. The domestication of grain began some 5000 years before and some enterprising farmer is bound to have become tired of pounding or grinding the grain by hand. Perhaps, in fact, the inventor was some farmer's wife, since she often drew the heavy jobs.

Many mills stayed in use until the end of the 19th century, but they had been in the process of being replaced by the steam engine as a source of power for the previous 150 years. The transmission of the power generated by the waterwheels was by shafting and crude gears or pulleys. The steam engine for the first time employed an enclosed moving stream of fluid (steam) under pressure—the principle to be later employed in hydraulic power transmission — for transmitting power continuously from the point of generation to the place where it was used.

The transmission of hydrostatic power to a distant point began long before that, however. Hero of Alexandria, in the first century AD, built a device in which a fire on the temple altar expanded air in a closed container. The air pressure forced water to travel along a tube to the temple doors where, spilling into a container, it provided the force through ropes and pulleys to magically open the doors.

Hero also produced a steam engine of sorts. Its motion depended on the reaction forces from jets of steam, as in the present steam turbine, but it was little understood and it was no more than an interesting curiosity.

The invention of early forms of the hydraulic pump has a similarly early origin. The first pumps were not used to develop hydraulic power, however, but only to transfer water for irrigation or to remove it from mines.

Archimedes applied the principle of the screw to hydraulic machines in the third century BC. His screw pumps were used to raise water for irrigation or to the level of aqueducts. The piston pump, the first mechanical device capable of generating pressure in a column of liquid, is believed to have come from Egypt at a similarly early date[②]. Like the screw pump, it was used only as a means for moving water and not as a means for generating hydraulic power.

In more recent years, the role of leadership in hydraulic power application has been taken over

largely by some of the larger earthmoving and construction equipment manufacturers③. The total power involved is often greater than that required in even the largest aircraft systems. The concentration of this power in a few very large loads, as compared to the multitude of smaller loads in an airplane, has spurred the system designer to find new, creative ways to distribute and control the power. This he has succeeded in doing to obtain maximum precision of control and productivity while minimizing power consumption.

In modern mobile and industrial applications, the system designer has made full use of the wonderful flexibility and adaptability of hydraulics. It can readily be directed around corners and past involving joints that virtually defy mechanical power transmission. It can be divided easily to serve individual loads or recombined to power a high, single-load demand. Electrical and pneumatic power distribution have similar flexibility, but lack the compactness and certain of the other important characteristics of hydraulics.

A very significant feature of hydraulic power is its extreme power density④. Fig. 10.1 shows a comparison of a 400 hp (1hp=0.745,7 kW) hydraulic pumping source with a diesel engine and an electrical motor of the same power capacity. The contrast is obvious and is one that offers both an opportunity to the user of hydraulic power and a challenge to the designer of hydraulic components.

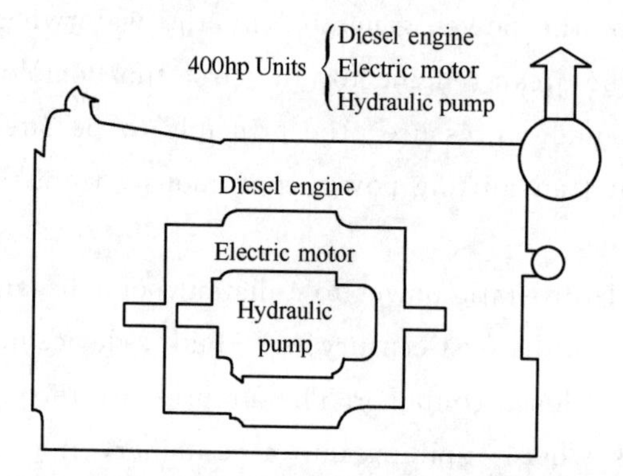

Fig. 10.1 Comparative size of diesel, electric and hydraulic power units

The high-power density in a hydraulic pump or motor creates design challenges beyond those found in many other products. High loads on rapidly sliding surfaces, extreme fluid velocities and rapidly recurring pressure applications on pumping elements (vanes, pistons, gears) demand careful design attention to assure long life, complete reliability and efficient operation. The user, on the other hand, reaps the benefit of a family of apparatus that is highly responsive, finely controllable and simple to apply. When the designer understands the user's needs and the user understands the equipment's full capabilities, the maximum potential of hydraulic power can be achieved.

Selected from "Hydraulic Pumps and Motors: Selection and Application for Hydraulic Power Control System", Raymond P. Lambeck, Marcel Dekker, Inc., 1983.

New Words and Expressions

1. prehistoric ['priːhis'tɔrik] *a*. 史前的，很久以前的

2. harness [ˈhɑːnis] *vt*. 利用（风等）作动力，治理，控制
3. watermill *n*. 水车，水磨
4. mosaic [məuˈzeiik] *n*. 镶嵌细工，马赛克
5. domestication [dəˌmestiˈkeiʃən] *n*. 家养，驯养
6. waterwheel *n*. 水轮，水车，辘轳
7. pulley [ˈpuli] *n*. 滑车，滑轮
8. altar [ˈɔːltə] *n*. 祭坛，圣坛
9. spill [spil] *v*.; *n*. 溢出，流出
10. aqueduct [ˈækwiˌdʌkt] *n*. 高架渠，渡槽
11. earthmoving [ˈəːθmuːviŋ] *a*. 大量撅土的，大量运土的
12. spur [spəː] *v*.; *n*. 刺激，激励，鼓励，推动
13. flexibility [ˌfleksəˈbiliti] *n*. 柔性，灵活性，柔韧性，
14. adaptability [əˌdæptəˈbiliti] *n*. 适应性
15. defy [diˈfai] *v*.; *n*. 使不能 [难以，落空]，向……挑战
16. compactness [kəmˈpæktnis] *n*. 紧密，紧密度，简洁，致密性
17. diesel [ˈdiːzəl] *n*. 柴油机

Notes

① 参考译文：水力的历史由来已久，始于人类为利用它周围的能源而做出的努力。

dating from…现在分词作状语。

② 参考译文：柱塞泵是历史上第一台能使液柱内产生压力的机械装置，它被认为来源于埃及且与此（螺旋泵）有着同样悠久的历史。

句中的 the first mechanical device capable of generating pressure in a column of liquid 作 the piston pump 的同位语。

③ 参考译文：近年来，一些规模较大的生产土建设备的厂商在液压动力应用方面一直占着主导地位。

这是一个被动语态的句子，主语为 the role of leadership in hydraulic power application，其中 in hydraulic power application 为介词短语作 the role of leadership 的定语；in more recent years 是介词短语，作全句的状语。

④ 参考译文：水力的一个明显的特点就是它具有极高的功率密度。

句子中的 power density 译为“功率密度”。

Exercises

1. After reading the text above，summarize the main ideas of it in oral English.
2. Answer the following questions according to the text.

   ① Please try to dictate the history of hydraulic system in your own words.

   ② What are the only sources readily available when man applied hydraulic power in the prehistory?

   ③ Who applied the principle of the screw to hydraulic machines in the third century BC?

   ④ What is the significant feature of the hydraulic power?
3. Translate the 8th paragraph into Chinese.
4. Put the following into Chinese by reference to the text.

hydraulic motor　heavy job　steam engine　enterprising farmer　flexibility
be employed in　multitude　screw pump

5. Put the following into English.
水力系统/液压系统　起始于……，溯源至……　史前的
土建设备　依赖于……　被……所取代

6. Translate the following sentences into English.
① 近年来，一些规模较大的生产土建设备的厂商处于液压动力应用的主导地位。
② 水利的一个很明显的特征是它具有极高的功率密度。

Reading Material 10

Valves

Process plants are a network of complex systems and processes. Just as arteries, veins, and the heart are vital to human life, pipes, valves, and pumps are indispensable in a process plant. The primary purpose of a valve is to direct and control the flow of fluids by starting, stopping, and throttling (restricting) flow to make processing possible. Valves are designed to withstand pressure, temperature, and flow and can be found in homes and industry across the world.

The common valves (Fig. 10. 2) found in the manufacturing environment are gate valves, globe valves, ball valves, check valves, butterfly valves, plug valves, needle valves, three-way valves, diaphragm valves, relief and safety valves, angle valves, and multiport valves. Valves normally are selected for a specific purpose. As you continue to read through this chapter, you will notice the variety of valve designs that exist. Operators need to be aware of how each valve works and the specific service for which it was designed.

Fig. 10. 2　Valves

Fig. 10. 3　Gate valve

Classification of Valves

Process operators classify valves by (1) flow-control elements, (2) function, and (3) operating conditions such as pressure, flow, or temperature. The most common way to classify valves is by the valve's flow-control element design. This part of the valve controls

or regulates the flow of fluid through the device. Some valves have movable metal gates, balls, plugs, diaphragms, discs, needles, or even butterfly-shaped elements. Most valves are named for the type or design of the flow-control element. Valves that are used for isolation are classified as block valves. While gate valves are the most common valve used for isolation, any valve can be used for this type of service. Another term associated with valve operation is valve capacity. Valve capacity is a term used to describe the total amount of fluid a valve will pass with a given pressure difference when it is fully open.

Gate Valves

One of the more common valves found in industry is a gate valve (Fig. 10. 3). A gate valve places a movable metal gate in the path of a process flow in a pipeline. The gates are sized to fit the inside diameter of a pipe, so very little restriction occurs when it is in the open position. Valves vary in size from 0. 125 inches to several feet. Gate valves typically are operated in the "wide open" or "completely shut" position. This type of valve is used where flow rates are not restricted. Gate valves should not be used to throttle flow for extended periods. Turbulent flow rates across the valve body will cause metal erosion, seat damage, and damage to the flow-control element, which can prevent the valve from blocking the flow completely.

The typical gate valve consists of a gate, body, seating area, stem, bonnet, packing, stuffing box, packing gland, and handwheel (Fig. 10. 4). The gate can be wedge shaped or may consist of parallel discs. It can be composed of a variety of materials. The gate is placed directly in the path of a process flow when it is shut and is lifted completely out of the way when open.

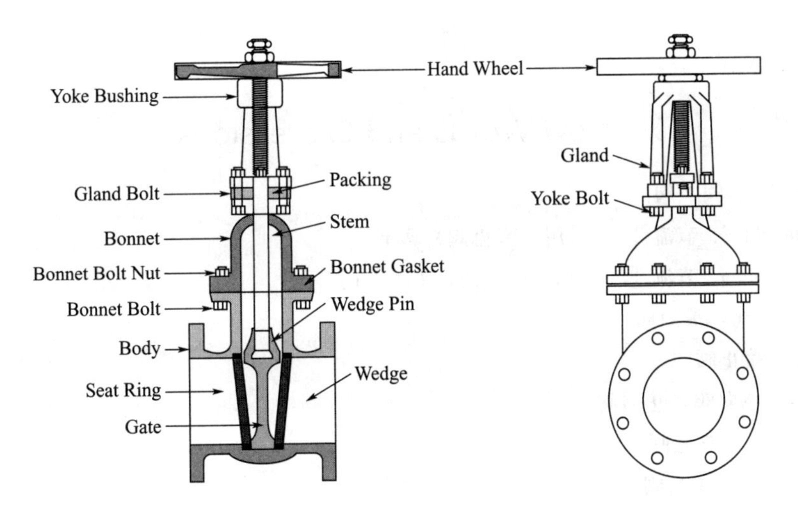

Fig. 10. 4 Gate valve components

The body is the largest part of the valve. The body can be connected to the process piping in three ways: flanges, threaded connections, or welding. The rest of the valve is attached to the body.

The seating area consists of two fixed surfaces or rings inside the body of the valve that the gate closes against to stop flow. The seating area falls into two categories: replaceable or fixed. Seats must provide a clean mating surface for the gate to seal properly. The seat can be fabricated or cast as part of the valve, press-fit, threaded, or welded into place. Note that high-temperature and high-pressure situations may require a combination of threading and welding.

The stem is a long, slender shaft attached to the gate, bushing, or wheel. When the handwheel is turned, it transmits rotational energy to the stem, causing it to rise to open or lower to close. The bonnet provides a housing for the gate or disc when it is lifted out of the process flow. It is attached to the body permanently by welding or temporarily by threading or bolts.

The packing is a specially designed material that prevents leakage from the bonnet, yet allows the stem to move up and down smoothly.

The stuffing box is typically located where the stem goes through the bonnet. The stuffing box is a recessed area specially designed to allow packing to be mounted around the stem.

The packing gland is a device used to compress and secure the packing material into the stuffing box. The packing gland nuts are designed to be evenly tightened by a technician to stop leaks.

The handwheel is attached to the valve stem. The handwheel transfers rotational energy to the stem. This rotational energy controls the movement of the flow-control element. Turning the handwheel clockwise closes the valve. When the handwheel is turned counterclockwise, it is opened.

Selected from "Process Technology: Equipment and Systems (third edition)", Charles E. Thomas, DELMAR CENGAGE Learning, 2011.

New Words and Expressions

1. valve [vælv] *n*. 阀，阀门
2. throttle ['θrɔtl] *v*. 节流［气］，用（节流阀）调节
   n. 节流［气］阀，风［油，主气］门
3. gate valve 闸阀，滑门阀
4. globe valve 截止阀
5. check valve 单向阀，止回阀
6. three-way valve 三通阀
7. diaphragm valve 隔膜阀
8. relief valve 安全［减压，卸压，保险］阀
9. block valve 隔断阀，截断阀
10. seat [siːt] *n*. 阀座
11. stem [stem] *n*. 阀杆，杆
12. bonnet ['bɔnit] *n*. 阀帽，阀盖

13. stuffing box 填（料）函，填料箱［盒］
14. packing gland 填料压盖，密封压盖
15. flange［flændʒ］*n*. 法兰（盘），凸缘
16. thread［θred］*n*. 螺纹［齿，丝，线］，线（状物）
17. disc［disk］*n*. 阀瓣，圆盘
18. nut［nʌt］*n*. 螺母，螺帽

Unit 11 • Thermodynamics

Using Thermodynamics

Engineers use principles drawn from thermodynamics and other engineering sciences, such as fluid mechanics and heat and mass transfer, to analyze and design things intended to meet human needs. The wide realm of application of these principles is suggested by Table 11. 1, which lists a few of the areas where engineering thermodynamics is important. Engineers seek to achieve improved designs and better performance, as measured by factors such as an increase in the output of some desired product, a reduced input of a scarce resource, a reduction in total costs, or a lesser environmental impact. The principles of engineering thermodynamics play an important part in achieving these goals.

Table 11. 1 Selected areas of application of engineering thermodynamics

Automobile engines, Turbines, Compressors and pumps,
Fossil- and nuclear-fueled power stations, Propulsion systems for aircraft and rockets,
Combustion systems, Cryogenic systems, gas separation and liquefaction,
Cooling of electronic equipment, Heating, ventilating, and air-conditioning systems (Vapor compression and absorption refrigeration, Heat pumps),
Alternative energy systems [Fuel cells, Thermoelectric and thermionic devices, Solar-activated heating, cooling, and power generation, Magnetohydrodynamic (MHD) converters, Geothermal systems, Ocean thermal, wave, and tidal power generation, Wind power]
Biomedical applications (Life-support systems, Artificial organs)

Defining Systems

An important step in any engineering analysis is to describe precisely what is being studied. In mechanics, if the motion of a body is to be determined, normally the first step is to define a free body and identify all the forces exerted on it by other bodies. Newton's second law of motion is then applied. In thermodynamics the term system is used to identify the subject of the analysis. Once the system is defined and the relevant interactions with other systems are identified, one or more physical laws or relations are applied①.

The system is whatever we want to study. It may be as simple as a free body or as complex as an entire chemical refinery. We may want to study a quantity of matter contained within a closed, rigid-walled tank, or we may want to consider something such as a gas pipeline through which matter flows. The composition of the matter inside the system may be fixed or may be changing through chemical or nuclear reactions. The shape or volume of the system being analyzed is not necessarily constant, as when a gas in a cylinder is compressed by a piston or a balloon is inflated.

Everything external to the system is considered to be part of the system's surroundings. The

system is distinguished from its surroundings by a specified boundary, which may be at rest or in motion. You will see that the interactions between a system and its surroundings, which take place across the boundary, play an important part in engineering thermodynamics. It is essential for the boundary to be delineated carefully before proceeding with any thermodynamic analysis. However, since the same physical phenomena often can be analyzed in terms of alternative choices of the system, boundary, and surroundings, the choice of a particular boundary defining a particular system is governed by the convenience it allows in the subsequent analysis.

Types of Systems

Two basic kinds of systems are distinguished in this book. These are referred to, respectively, as closed systems and control volumes. A closed system refers to a fixed quantity of matter, whereas a control volume is a region of space through which mass may flow.

A closed system is defined when a particular quantity of matter is under study. A closed system always contains the same matter. There can be transfer of mass across its boundary. A special type of closed system that does not interact in any way with its surroundings is called an isolated system.

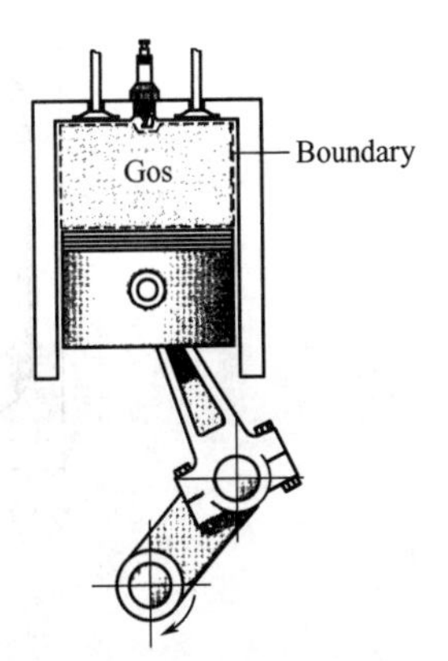

Fig. 11. 1 Closed system: A gas in a piston-cylinder assembly

Fig. 11. 1 shows a gas in a piston-cylinder assembly. When the valves are closed, we can consider the gas to be a closed system. The boundary lies just inside the piston and cylinder walls, as shown by the dashed lines on the figure. The portion of the boundary between the gas and the piston moves with the piston. No mass would cross this or any other part of the boundary.

In subsequent sections, thermodynamics analyses are made of devices such as turbines and pumps through which mass flows. These analyses can be conducted in principle by studying a particular quantity of matter, a closed system, as it passes through the device. In most cases it is simpler to think instead in terms of a given region of space through which mass flows. With this approach, a region within a prescribed boundary is studied. The region is called a control volume. Mass may cross the boundary of a control volume.

A diagram of an engine is shown in Fig. 11. 2(a) . The dashed line defines a control volume that surrounds the engine. Observe that air, fuel, and exhaust gases cross the boundary. A schematic such as in Fig. 11. 2(b) often suffices for engineering analysis.

The term control mass is sometimes used in place of closed system, and the term open system is used interchangeably with control volume. When the terms control mass and control volume are used, the system boundary is often referred to as a control surface.

In general, the choice of system boundary is governed by two considerations: (1) what is known about a possible system, particularly at its boundaries, and (2) the objective of the analysis. For example, Fig. 11. 3 shows a sketch of an air compressor connected to a storage tank. The system boundary shown on the figure encloses the compressor, tank, and

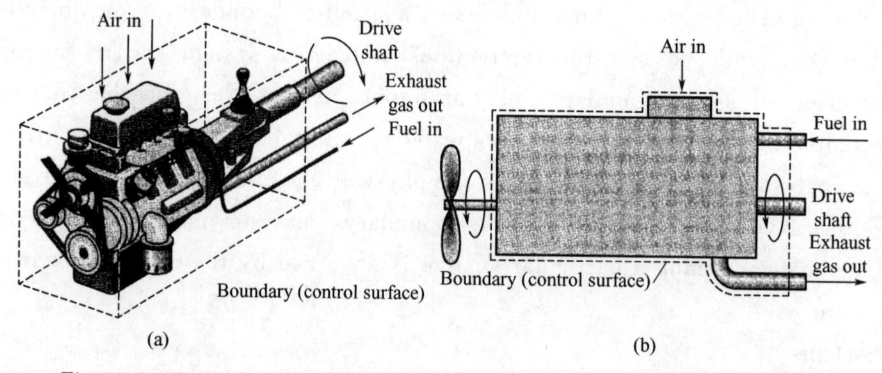

Fig. 11. 2　Example of a control volume (open system). A automobile engine

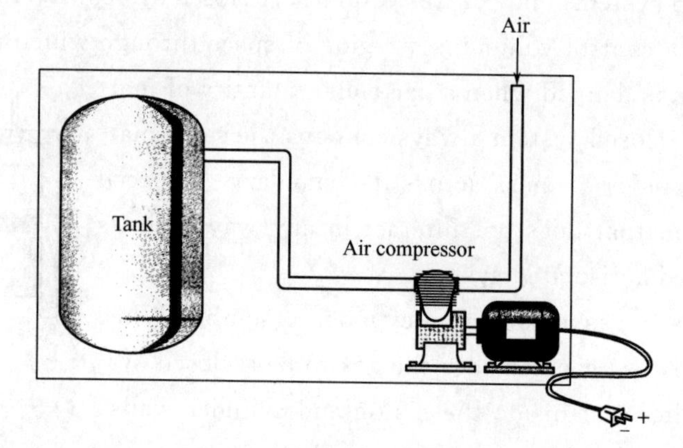

Fig. 11. 3　Air compressor and storage tank

all of the piping. This boundary might be selected if the electrical power input were known, and the objective of the analysis were to determine how long the compressor must operate for the pressure in the tank to rise to a specified value. Since mass crosses the boundary, the system would be a control volume. A control volume enclosing only the compressor might be chosen if the condition of the air entering and exiting the compressor were known, and the objective were to determine the electric power input.

Selected from "Fundamentals of Engineering Thermodynamics (6th Edition)", Michael J. Moran, Howord M. Shapiro, John Wiley & Sons Inc., 2010.

Words and Expressions

1. thermodynamics [θəːməudaiˈnæmiks] *n.* [物] 热力学
2. turbine [ˈtəːbin] *n.* 涡轮（机），透平（机），汽轮机
3. ventilate [ˈventileit] *v.* （使）通风［气］，排气
4. interaction [intərˈækʃən] *n.* 互相作用，互相影响，交互作用，互动
5. cylinder [ˈsilində] *n.* 圆筒，圆柱体，汽缸，柱面
6. inflated [inˈfleitid] *a.* 膨胀的，夸张的，通货膨胀的
7. delineate [diˈlineit] *vt.* 描绘……的轮廓，描绘，描写
8. subsequent [ˈsʌbsikwənt] *a.* 后来的，接下去的
9. dashed [dæʃt] *a.* 虚（线）的

10. schematic [ski'mætik] *n.*; *a.* 图表，示意性的

Notes

① 参考译文：一旦确定了这一系统，同时认为它与其它系统有相应互动关系，则可应用一个或多个物理定律或关系式。

Exercises

1. After reading the text above, summarize the main ideas of it in oral English.
2. Answer the following questions according to the text.
   ① What is the indispensable step in any engineering analysis?
   ② What is a free body (Please take an example to illustrate it)?
   ③ How is the system distinguished from its surroundings in thermodynamic analysis?
   ④ What is the difference between two basic kinds of systems, closed systems and control volumes?
3. Translate the 2nd paragraph into Chinese.
4. Put the following into Chinese by reference to the text.
   Newton's second law of motion　relevant interactions　control volume
   be distinguished from　control mass　surroundings　boundary
   air-conditioning system　thermoelectric and thermionic devices
5. Put the following into English.
   热力学第一定律　内能　废气　自由体，隔离体　工程热力学　热泵　燃料电池
6. Put the following sentences into English.
   ① 工程师用热力学和其它工程科学原理分析事物，从而满足人类的需要。
   ② 热力学系统中的研究对象很广泛，可能只是一个简单的自由体，也可能是复杂的整个炼油厂。

Reading Material 11

Applications of Engineering Thermodynamics

In the course of our study of thermodynamics, a number of the examples and problems presented refer to processes that occur in equipment such as a steam power plant, a fuel cell, a vapor-compression refrigerator, a thermoelectric cooler, a turbine or rocket engine, and an air separation plant. In this introductory chapter, a brief description of this equipment is given. There are at least two reasons for including such a chapter. First, many students have had limited contact with such equipment, and the solution of problems will be more meaningful when they have some familiarity with the actual processes and the equipment. Second, this chapter will provide an introduction to thermodynamics, including the use of certain terms, some of the problems to which thermodynamics can be applied, and some of the things that have been accomplished, at least in part, from the application of thermodynamics.

Thermodynamics is relevant to many other processes than those cited in this chapter. It is basic to the study of materials, chemical reactions, and plasmas. The student should bear in mind that this chapter is only a brief and necessarily very incomplete introduction to the

subject of thermodynamics.

The Simple Steam Power Plant

A schematic diagram of a recently-installed steam power plant is shown in Fig. 11. 4. High-pressure superheated steam leaves the steam drum at the top of the boiler, also referred to as a steam generator, and enters the turbine. The steam expands in the turbine and, in doing so does work, which enables the turbine to drive the electric generator. The steam, now at low pressure, exits the turbine and enters the heat exchanger, where heat is transferred from the steam (causing it to condense) to the cooling water. Since large quantities of cooling water are required, power plants have traditionally been located near rivers or lakes, leading to thermal pollution of those water supplies. More recently condenser cooling water is recycled by evaporating a fraction of the water in large cooling towers, thereby cooling the remainder of the water that remains as a liquid. In the power plant shown in Fig. 11. 4, the plant is designed to recycle the condenser cooling water by using the heated water for district space heating.

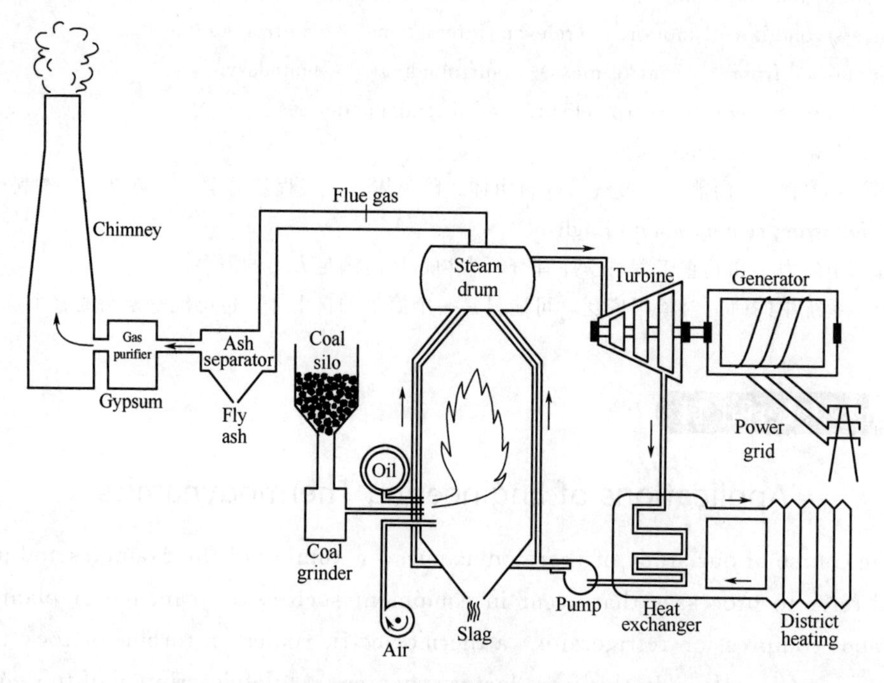

Fig. 11. 4 Schematic diagram of a steam power plant

The pressure of the condensate leaving the condenser is increased in the pump, enabling it to return to the steam generator for reuse. In many cases, an economizer or water preheater is used in the steam cycle, and in many power plants, the air that is used for combustion of the fuel may be preheated by the exhaust combustion-product gases. These exhaust gases must also be purified before being discharged to the atmosphere, such that there are many complications to the simple cycle.

The Vapor-Compression Refrigeration Cycle

A simple vapor-compression refrigeration cycle is shown schematically in Fig. 11. 5. The

refrigerant enters the compressor as a slightly superheated vapor at a low pressure. It then leaves the compressor and enters the condenser as a vapor at some elevated pressure, where the refrigerant is condensed as heat is transferred to cooling water or to the surroundings. The refrigerant then leaves the condenser as a high-pressure liquid. The pressure of the liquid is decreased as it flows through the expansion valve, and as a result, some of the liquid flashes into cold vapor. The remaining liquid, now at a low pressure and temperature, is vaporized in the evaporator as heat is transferred from the refrigerated space. This vapor then reenters the compressor.

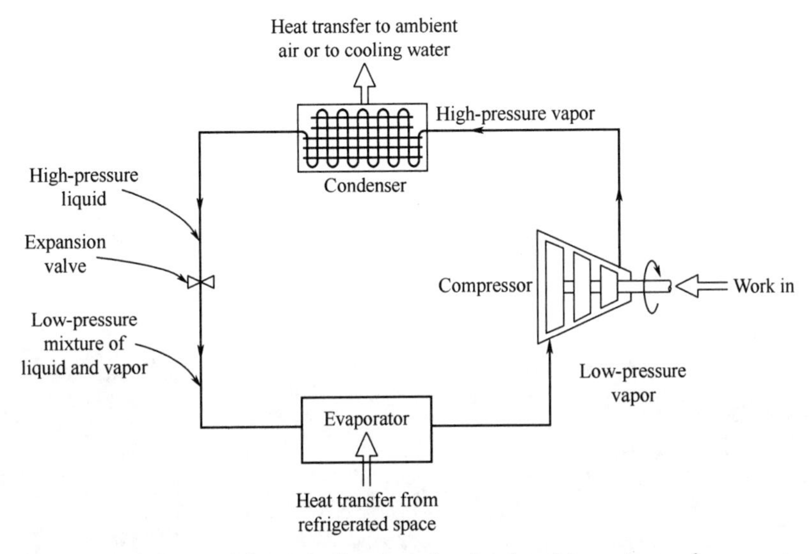

Fig. 11. 5 Schematic diagram of a simple refrigeration cycle

In a typical home refrigerator the compressor is located in the rear near the bottom of the unit. The compressors are usually hermetically sealed, that is, the motor and compressor are mounted in a sealed housing, and the electric leads for the motor pass through this housing. This seal prevents leakage of the refrigerant. The condenser is also located at the back of the refrigerator and is arranged so that the air in the room flows past the condenser by natural convection. The expansion valve takes the form of a long capillary tube, and the evaporator is located around the outside of the freezing compartment inside the refrigerator.

Selected from "Introduction to Engineering Thermodynamics", Richard E. Sonntag, Claus Borgnakke, John Wiley & Sons Inc., 2001.

Words and Expressions

1. power plant ［动力］发电厂，动力设备［装置］
2. refrigerator [ri'fridʒəreitə] *n*. 制冷器［机］，冷气机，（电）冰箱
3. engine ['endʒin] *n*. 发动机，引擎，火车头，机车
4. steam drum 蒸汽锅筒，上汽锅
5. generator ['dʒenəreitə] *n*. （蒸汽）发生器，发电［动］机
6. condense [kən'dens] *v*. 冷凝，凝结；浓［凝］缩

7. cooling tower 冷却塔
8. space heating 环流供暖
9. economizer [i'kɔnəmaizə] *n.* 省煤［油］器，废气预［节］热器
10. expansion valve 膨胀［安全，调节］阀
11. evaporator [i'væpəreitə] *n.* 蒸发［汽化］器
12. seal [siːl] *v.*；*n.* 密封
13. lead [liːd] *n.*（导，引）线，（电）线头
14. leakage ['liːkidʒ] *n.* 泄漏，渗漏

Unit 12 • Fluid Mechanics in Engineering

Fluid Mechanics is the branch of engineering science that is concerned with forces and energies generated by fluids at rest and in motion. The study of fluid mechanics involves application of the fundamental principles of mechanics and, to a lesser degree, of thermodynamics to develop a qualitative understanding and quantitative analysis techniques that an engineer can apply to design or evaluation of equipment and processes that involve fluids①.

You are probably reading this book because fluid mechanics is a required course in your curriculum. The study of fluid mechanics is included in most engineering curricula because the principles and methods of fluid mechanics find many technological applications. To illustrate a few applications, we will consider the following fields:

① Fluid transport;

② Energy generation;

③ Environmental control;

④ Transportation.

Fluid transport is movement of a fluid from one place to another so that the fluid may be used or processed. Examples include home and city water supply systems, cross-country oil, natural gas, and agricultural chemical pipelines, and chemical plant piping. Engineers involved in fluid transport might design systems involving pumps, compressors, pipes, valves, and a host of other components all directly involved with motion of fluids. In addition to the design of new systems, engineers may evaluate the adequacy of existing systems to meet new demands or they may maintain or upgrade existing systems.

In the field of energy generation, we find that, with the exception of chemical batteries and direct energy-conversion devices such as solar cells, no useful energy is generated without fluid movement. Typical energy-conversion devices such as steam turbines, reciprocating engines, gas turbines, hydroelectric plants, and windmills involve many complicated flow processes. In all of these devices, energy is extracted from a fluid, usually in motion. The thermodynamic cycles of these devices usually require fluid machinery such as pumps or compressors to do work on the fluid. Auxiliary equipment such as oil pumps, carburetors, fuel-injection systems, boiler, draft fans, and cooling systems also depends on fluid motion.

Environmental control involves fluid motion. An estimated 75 percent of American homes are heated by forced-air systems, in which the home air is continually recirculated to transport heat from the combustion of fuel. In air-conditioning systems the circulating air is cooled by a flowing refrigerant. Similar processes occur in automobile engine cooling systems, in machine tool cooling systems, and in the cooling of electronic components by ambient air.

With the exception of space travel, all transportation takes place within a fluid medium (the atmosphere or a body of water). The relative motion between the fluid and the transportation device is responsible for the generation of a force that opposes the desired motion. This force can be minimized by the application of fluid mechanics to vehicle design. The fluid often contributes in a positive way, such as by floating a ship or generating lift by air motion over airplane wings. In addition, ships and airplanes derive propulsive force from propellers or jet engines that interact with the surrounding fluid.

These examples are by no means exhaustive. Other engineering applications of fluid mechanics include the design of canals and harbors as well as dams for flood control. The design of large structures must account for the effects of wind loading. In the relatively new fields of environmental engineering and biomedical engineering, engineers must deal with naturally occurring flow processes in the atmosphere and lakes, rivers, and seas or within the human body. Although it is not usually considered an engineering discipline, the phenomena of fluid motion are central to the field of meteorology and weather forecasting.

Few engineers can function effectively without at least a rudimentary knowledge of fluid mechanics. Large numbers of engineers are primarily involved with processes, devices, and systems in which a knowledge of fluid mechanics is essential to intelligent design, evaluation, maintenance, or decision making②. You probably cannot foresee exactly what problems you will be called on to solve in your professional career; therefore, you would be wise to obtain a firm grasp of the fundamentals of fluid mechanics.

These fundamentals include a knowledge of the nature of fluids and the properties used to describe them, the physical laws that govern fluid behavior, the ways in which these laws may be cast into mathematical form, and the various methodologies (both analytical and experimental) that may be used to solve engineering problems.

Selected from "Fundamentals of Fluid Mechanics", Philip M. Gerhart, Richard J. Gross, Addison-Wesley Publishing Company, 1985.

New Words and Expressions

1. environmental [inˌvaiərənˈmentəl] *a*. 环境的，环境产生的
2. hydroelectric [ˈhaidrəiˈlektrik] *a*. 水力电气的，水电的
3. carburetor [kɑːbəˈretə] *n*. 汽化器
4. refrigerant [riˈfridʒərənt] *a*.; *n*. 制冷的；制冷剂
5. propulsive [prəuˈpʌlsiv] *a*. 推进的，有推进力的
6. propeller [prəˈpelə] *n*. 推进器，螺旋桨
7. exhaustive [igˈzɔːstiv] *a*. 无遗漏的，彻底的，详尽的
8. dam [dæm] *n*. 堤，坝
9. biomedical [ˌbaiəuˈmedikəl] *a*. 生物医学的
10. meteorology [miːtjəˈrɔlədʒi] *n*. 气象学，气象状态
11. rudimentary [ruːdiˈmentəri] *a*. 根本的，未发展的，基本的，初步的
12. foresee [fɔːˈsiː] *vt*. 预见，预知
13. methodology [meθəˈdɔlədʒi] *n*. 方法学，方法论

Notes

① 参考译文：流体力学的研究涉及力学基本原理的应用和较少热力学基本原理的应用，用以发展一种定性认知和定量分析的技术，使工程师能用于设计或评估与流体有关的各种设备与过程。

此句为简单句。主语是 the study of fluid mechanics，involves 是谓语，application of … fluids 是宾语。句中有三个连词 and，第一个连接 of mechanics 和 of thermodynamics，并列作 the fundamental principles 的定语。第二个连接 qualitative understanding 和 quantitative analysis，修饰 techniques。第三个连接 equipment 和 processes，并列作 design or evaluation 的定语。两个由 that 引导的定语从句分别修饰 techniques 和 equipment and processes。

② 参考译文：很多工程师主要从事过程、装置或系统的工作，流体力学是进行智能设计、评估、维护和决策不可缺少的知识。

Exercises

1. After reading the text above，summarize the main idea of it in oral English.
2. Answer the following questions according to the text.
   ① What is the definition of fluid mechanics according to the text?
   ② Which applications are involved during the study of fluid mechanics?
   ③ Can you try to make a list of typical energy-conversion devices after reading the text?
   ④ What knowledge do the fundamentals of fluid mechanics include?
3. Translate the 1st and 8th paragraphs into Chinese.
4. Put the following into Chinese by reference to the text.

be concerned with	fluid transport	energy-conversion devices	a host of
fluid machinery	auxiliary equipment	air-conditioning systems	cast into

5. Put the following into English.

流体力学	定性认知	供水系统	能量损失
必修课	定量分析	流率	

6. Put the following sentences into English.
   ① 流体力学是一门流体在静止和运动时产生力和能量的工程科学。
   ② 流体输送是指为利用和处理流体而使之从一个地方到另一个地方的运动过程。

Reading Material 12

Regimes of Flow in a Straight Pipe or Duct

In the text we discuss or analyze flow in straight pipes and ducts. Before we can discuss or analyze the flow in a particular pipe，we must know if the flow is laminar or turbulent and if it is developing or fully developed.

Laminar Flow and Turbulent Flow

If the flow in a pipe is laminar，the fluid moves along smooth streamlines. The fluid velocity at any point in the pipe is virtually constant in time. If the flow is turbulent，a rather

violent mixing of the fluid occurs, and the fluid velocity at a point varies randomly with time.

The differences between laminar and turbulent pipe flows were first clarified by Osborne Reynolds in 1883. Reynolds performed a series of experiments in which dye was injected into water flowing in a glass pipe. At low flow rates, the dye streak remained smooth and regular as the dye flowed downstream. At higher flow rates, the dye seemed to explode, mixing rapidly throughout the entire pipe. A modern spark photograph of the mixing dye would reveal a very complex flow pattern, not discernible by Reynolds's experiments.

Reynolds's experiments and his analytical work showed that the nature of pipe flow depends on the Reynolds number

$$R=\frac{\rho VD}{\mu} \tag{12-1}$$

Flow in Pipe and Duct Systems

Pipe and duct systems include many components in addition to pipes and ducts. We will now consider the flow in some of these components. As in our analysis of pipe/duct flow, our primary objective is to discuss relations between the component geometry, the flow rate through the component, and the mechanical energy loss generated by the component. After considering the components, we will examine complete systems made up by combining pipes or ducts with various components.

System Components and Local Losses

Some auxiliary components that are used in a pipe or duct system cause turbulence and mechanical energy loss in addition to that which occurs in the basic pipe flow. Fig. 12. 1 illustrates flow through a sudden enlargement, a rather crude transition. The flow is not able to turn the sharp corner, so it separates, generating a great deal of disorderly motion. The disorderly motion is often loosely called "turbulence"; however, the motion near the sudden

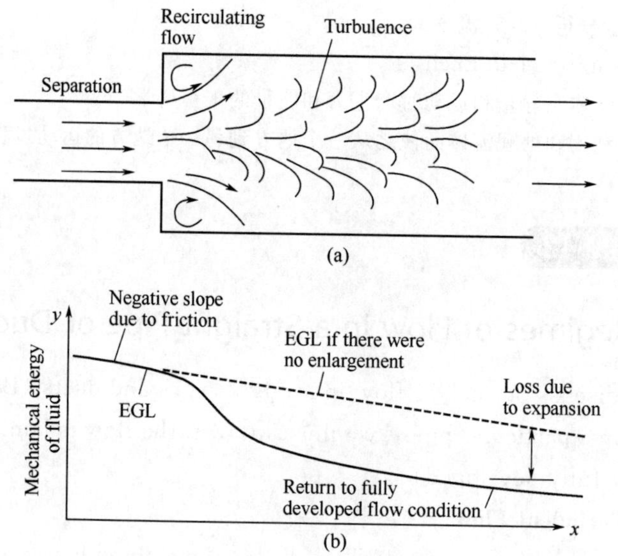

Fig. 12. 1　Flow detail energy loss for flow through a sudden enlargement

enlargement is more orderly than true turbulence. The motion degenerates into true turbulence as the fluid proceeds downstream. The disturbance caused by the expansion persists for some distance downstream as the kinetic energy of the disorderly motion is dissipated and the flow gradually returns to a fully developed condition. This flow behavior is not unique to the sudden expansion; other components generate similar disturbances. These disturbances, loosely called "turbulence" are responsible for energy loss in the region immediately downstream of the component. More profound disturbances persist further downstream and produce larger losses. A plot of the energy grade line (EGL) in the immediate vicinity of the component shows that the energy loss occurs over a finite distance; however, when viewed from the perspective of an entire pipe system, the energy losses are localized near the component. We will refer to such losses as local losses.

Most authors and many engineers refer to such losses as minor losses. This is an unfortunate misnomer because local losses may be the dominant losses in some systems. Local losses are added to pipe friction losses to calculate the total energy loss in the system. Local loss is usually the total loss in the component. The length of the component should not be included in the pipe length when pipe friction losses are calculated.

Local losses are calculated by using a loss coefficient K

$$gh_{\mathrm{L}}=K\left(\frac{V^2}{2}\right) \tag{12-2}$$

The loss coefficient is a function of component geometry and Reynolds number. A scale-effect parameter similar to relative roughness in pipe flows may also be important. Since most components generate energy loss by promoting turbulence, loss coefficients are usually nearly independent of Reynolds number.

An alternative to the loss coefficient method of calculating local losses is the "equivalent length" method. In this method, the component is "replaced" by a straight run of pipe (duct) that would have the same loss. Equating the equivalent friction loss to local loss, we have

$$f\frac{L_{\mathrm{eq}}}{D_{\mathrm{h}}}\frac{V^2}{2}=K\frac{V^2}{2} \tag{12-3}$$

So the equivalent length is calculated by

$$L_{\mathrm{eq}}=\frac{K}{f}D_{\mathrm{h}} \tag{12-4}$$

Where f is the friction factor in the pipe (duct) to which the component is attached. The equivalent length concept is reasonably straight forward for all components except transitions. For transitions, one must be careful to specify whether the equivalent length is added to the smaller or larger pipe (duct).

Selected from "Fundamentals of Fluid Mechanics", Philip M. Gerhart, Richard J. Gross, Addison-Wesley Publishing Company, 1985.

New Words and Expressions

1. duct [dʌkt] *n.* 管，输送管，排泄管（指非圆形的）

2. laminar ['læminə(r)] *a*. 由薄片或层状体组成的，薄片状的，层流的
3. turbulent ['tə:bjulənt] *a*. 湍动的，湍流的
4. randomly [rændəmli] *ad*. 随机地，随便地，未加计划
5. turbulence ['tə:bjuləns] *n*. 骚乱，动荡，湍流或紊流
6. inject [in'dʒekt] *n*. 注射，注入
7. downstream ['daunstri:m] *a*. 下游的
8. dominant ['dɔminənt] *a*. 占优势的，支配的，有统治权的
9. profound [prə'faund] *a*. 深刻的，意义深远的，渊博的
10. coefficient [kəui'fiʃənt] *n*. [数] 系数

Unit 13 ● Compressors —General Type Selection Factors

Compressors are used in petrochemical plants to raise the static pressure of air and process gases to levels required to overcome pipe friction, effect a certain reaction at the point of final delivery, or to impart desired thermodynamic properties to the medium compressed. These compressors come in a variety of sizes, types, and models, each of which fulfills a given need and is likely to represent the optimum configuration for a given set of requirements①.

Compressor type selection must, therefore, be preceded by a comparison between service requirements and compressor capabilities. This initial comparison will generally lead to a review of the economies of space, installed cost, operating cost, and maintenance requirements of competing types. Where the superiority of one compressor type or model over a competing offer is not obvious, a more detailed analysis may be justified.

Compression machinery can be separated into two broad categories: dynamic and positive displacement②. The centrifugal compressor is a dynamic machine by contrast to the static, positive-displacement type of compressor. Centrifugal compressors do their work by using inertial forces applied to the gas by means of rotating bladed impellers, whereas positive-displacement compressors trap gas by the action of mechanical components and restrict its escape as compression takes place through direct volume reduction. Each of these two broad categories can be further subdivided as shown in Fig. 13. 1(a) & (b).

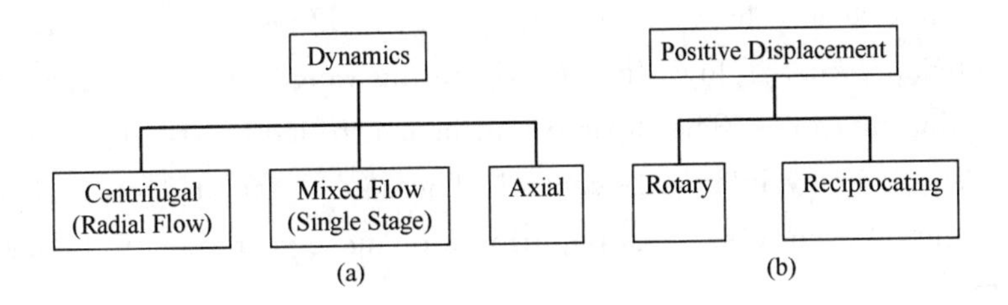

Fig. 13. 1 Chart of principal compressor types

A number of service conditions affect the type selection. Foremost among these are volume flow rate and discharge pressure. Both of these parameters, in turn, influence power levels.

Aside from the limitations of flow and discharge pressure, compressor type selection will be affected by the characteristics of the process system. Each of the different types of compressors has characteristics which make it more suitable for some application than for another. Some of these features are listed below.

Reciprocating Compressors

Reciprocating compressors are available for almost all compressor applications. They are suitable for all pressures from vacuum to around 100,000 lb/in^2 （1 lb/in^2 =6.894,76×10^{-3} MPa） gauge. They are available to handle volumes from less than 5 ACFM up to 7,000 ACFM. Their overall efficiency varies from 80% to 90%， averaging about 85%.

The disadvantages of these machines are: For continuous duties such as powerformer hydrogen recycle compression， more than one machine must be provided to permit servicing. They are large and expensive. They have a high maintenance cost， especially when handling gases containing liquids， solids， or corrosive materials. Because they have large shaking forces， large foundations are required.

Rotary Screw Compressors

This compressor type is generally available for pressures up to 250 lb/in^2 gauge and for volumes of 800 to 20,000 ACFM. They are balanced machines and require only light foundations. Because there are no rubbing surfaces， they do not contaminate the compressed gas with lubricating oil. The maintenance cost should be low. In their operating range they are cheaper than both centrifugal and reciprocating machines. Their efficiency is in the range of 75% to 85%. They can be used to compress gases containing tars or polymers， and in these cases the efficiency of the machine is higher than when handling clean gases.

The disadvantages of these machines are: The machines are noisy. The range of capacity variation possible at constant speed is very small. The machines are designed for a specified gas and compression ratio.

Centrifugal Compressors

Centrifugal compressors are extensively used in modern petrochemical plants. They are basically large volume machines. They are available for pressures of up to over 5000 lb/in^2 gauge and handle volumes of 1,000 to 150,000 ACFM. Because there are no rubbing surfaces， they do not contaminate the compressed gas with lubricating oil. They are balanced machines and do not require heavy foundations. Their efficiency is in the range of 68% to 76%. The maintenance cost is low. In their operating range， their initial cost is less than that of reciprocating machines. The capacity can be controlled by speed variations， reducing the suction pressure， or by inlet vane control. The service factor is so high that only one compressor is required even in services requiring 3 or more years in continuous operation.

Axial Flow Compressors

Axial flow compressors are essentially very high volume machines. Except at volume flows over 60,000 ACFM， they are more costly than centrifugal compressors. The efficiency of an axial compressor is as much as 10% higher than it is for a centrifugal compressor. Discharge pressures of 200 lb/in^2 gauge have been demonstrated on axial flow process machines.

Selected from "Compressors and Expanders， Selection and Application for the Process Industry"， Heinz P. Bloch， Joseph A. Cameron， *et al.*， Marcel Dekker， Inc.， 1982.

New Words and Expressions

1. **compressor** [kəm'presə] *n*. 压缩机

2. petrochemical [ˌpetrəuˈkemik(ə)l] *a.*; *n.* 石化的；石化产品
3. delivery [diˈlivəri] *n.* 输出，交货
4. thermodynamic [ˌθəːməudaiˈnæmik] *a.* 热力学的，使用热动力的
5. precede [pri(ː)ˈsiːd] *v.* 在……之前，优于，较……优先
6. superiority [sju(ː)piəriˈɔriti] *n.* 优越性
7. bladed impeller 装有叶片的叶轮
8. subdivide [ˈsʌbdiˈvaid] *v.* [把……] 再分，[把……] 细分
9. foremost [ˈfɔːməust] *a.*; *ad.* 最重要的；在最前
10. ACFM=absolute cubic feet per minute 绝对立方英尺/分
11. corrosive [kəˈrəusiv] *a.* 腐蚀的，腐蚀性的
12. rotary screw compressor 回转式螺旋压缩机
13. rubbing [ˈrʌbiŋ] *n.* 摩擦，研磨
14. contaminate [kənˈtæmineit] *vt.* 污染，弄脏
15. tar [tɑː] *n.* 焦油
16. vane [vein] *n.* 叶片

Notes

① 参考译文：这些压缩机有各种尺寸、类型和型号，其中每一种都能满足一已知需求，而且对于给定的一组要求而言，有可能是最优配置。

② 参考译文：压缩机类机械可以分成两大类：动力类与容积类。

dynamic compressors 动力压缩机。动力压缩机有：

离心式压缩机　　centrifugal compressor；
轴流式压缩机　　axial-flow compressor；
混流式压缩机　　mixed-flow compressor.

positive displacement compressors 变容式压缩机，习惯上称“容积式压缩机”。容积式压缩机可分成两大类：往复式结构（reciprocating）和回转式结构（rotary）。回转式结构常见的有：

滑片式压缩机　　sliding-vane compressor；
液环式压缩机　　liquid-piston compressor；
罗茨压缩机　　two-impeller straight-lobe compressor；
螺杆式压缩机　　helical- or spiral-lobe compressor.

Exercises

1. After reading the text above, write a summary of it.
2. Answer the following questions according to the text.
   ① Please try to describe the main use of compressors and the construction of them in your own words.
   ② What initial comparison should be made between different types when the compressor is chosen?
   ③ Which two categories can compressors be classified into?
   ④ Please try to make a comparison between centrifugal compressors and axial flow compressors.
3. Translate the 6th paragraph into Chinese.

4. Put the following into Chinese by reference to the text.

   by contrast to   discharge pressure   volume flow rate   reciprocating machine

   axial flow compressor   be preceded by

5. Put the following into English.

   离心泵   润滑油   压缩比   惯性力   维修费用

6. Translate the following sentences into English.

   ① 压缩机被广泛应用于石油化工行业，通过它来克服管道阻力，提高空气和工作气体的压力，使之达到要求的标准。

   ② 选择压缩机类型，除了受到流速和压力的限制以外，还受到过程系统特性的约束。

Reading Material 13

Pumps

Fluids are moved through pipe, equipment or the ambient by pumps, fans, blowers and compressors. These machines are not only used to transfer the fluids but also used to add energy to the fluids. Such devices increase the mechanical energy of the fluid. The energy increase may be used to increase the velocity, the pressure, or the elevation of the fluid. Among these pumps are machines used to transfer liquids and add energy to the liquids.

Pumps are used in a variety of applications and processes including refrigeration, automobiles, home heating systems, and water wells. The liquids moved by a pump vary from liquid sodium and liquid potassium for cooling nuclear reactors to domestic drinking water system.

In general, pumps can be classified as dynamic or positive displacement. Both classes are designed to transfer liquids, but the way the transfer is accomplished is different.

Dynamic pumps include centrifugal and axial types. They operate by developing a high liquid velocity, which is converted to pressure in a diffusing flow passage. These pumps generally are lower in efficiency than the positive displacement types. However they do operate at relatively high speeds, thus providing high flowrate in relation to the physical size of the pump. Furthermore, they usually have significantly lower maintenance requirements than positive displacement pumps.

Positive displacement pumps operate by forcing a fixed volume of fluid from the inlet pressure section of the pump into the pump's discharge zone. This is performed intermittently with reciprocating pumps. In the case of rotary screw and gear pumps, the action is continuous. This category of pumps operates at lower rotating speeds than does dynamic pump. Positive displacement pumps also tend to be physically larger than equal-capacity dynamic pumps.

Dynamic pumps are classified as centrifugal or axial. Centrifugal pumps operate on the principle of centrifugal force. Centrifugal pumps are used extensively throughout the process industries because of their simplicity in design, low initial cost, low maintenance, less space, flexibility of application, and they will operate with a constant head pressure over a

wide capacity range. A spinning impeller inside a shell casing propels liquid outward. Fluid velocity is accelerated inside the shell of the pumps and liquid quickly moves toward the discharge port.

The axial pumps utilize a similar spinning motion to propel liquid, but the liquid moves in a straight line. This motion is directionally different from the centrifugal pumps' outward movement.

Rotary and reciprocating pumps are the two major classifications of positive displacement pumps. Rotary pumps displace liquid with rotary-motion gears, screws, vanes, or lobes. Reciprocating pumps displace fluid with a diaphragm, piston, or plunger that moves back and forth.

Rotary pumps are positive displacement pumps that transfer liquids using a rotary motion. The drive shaft turns the rotary element inside a leaktight chamber that has a defined inlet and outlet. Rotary pumps are used to move the more viscous type of fluids: heavy hydrocarbons, syrup, paint, and slurries. Rotary pumps combine the rotary motion of a centrifugal pump and the positive displacement feature of a reciprocating pump. Rotary pumps come in four main types: screw (single and multiple), gear (internal and external), vane (sliding and flexible), and lobe.

This is a single-screw rotary pump, just has one screw. But multiple screw rotary pumps can be used in a variety of application. They have either two or three screws.

Gear pumps are similar to screw pumps in that they can be used in viscous service. Gear pumps typically can be found in two common types: external and internal.

External gear pumps have two interesting gears that rotate parallel to each other, allowing fluid to be picked up by the gears and transferred out of the pump.

Internal gear pumps operate with only two moving parts: a power gear driving an internal idler gear. When the power gear rotates, liquid enters the pump through the suction line.

Sliding vane pumps consist of spring-loaded or non spring loaded vanes attached to a rotor, or impeller, that rotates inside an oversized circular casing. As the offset impeller rotates by the inlet port, liquid is swept into the vane slots. A small crescent-shaped cavity is formed inside the pumping chamber that the vanes extend into. As the liquid nears the discharge port, it is compressed as the clearances narrow. The compressed liquid is released at the discharge port. The vanes on the pump are made of a softer material than the rotor and casing. Vane pumps are used in hydraulic systems, vacuum systems, and low-pressure oil systems.

In a flexible pump system the rotor is composed of a soft elastomer impeller, keyed to fit over the drive shaft that penetrates the pumping chamber. The pumping chamber is designed to provide good contact between the impeller and the inner chamber. Flexible vane pumps are frequently used in vacuum service.

Lobe pumps have two rotating lobe-shaped screws that mesh during operation. As the lobes turn, voids are created that compress liquids around the outside of the pumping chamber.

Lobe pumps are designed to provide high flowrates at low pressures; they have excellent suction and pump a variety of fluids.

Reciprocating pumps, especially in small volume sizes, are positive displacement pumps very commonly used in the petrochemical industry. Reciprocating pumps are engineered to transfer small volumes of liquid at relatively high pressures. Most reciprocating pumps are self-priming and are operated at relatively low speeds because of the back-and-forth motion and the effects of inertia on internal components. Reciprocating pumps can be deliver consistently high volumetric efficiencies even when applied to a variety of fluid types. They deliver liquid when a piston, plunger, or diaphragm physically displaces it. The diaphragm, piston, or plunger pushes the fluid as it moves back and forth inside a cylinder or housing.

Four characteristics describe all pumps:

① Capacity $Q(m^3/s)$—the quantity of liquid discharged per unit time;

② Head $H(m)$—the energy supplied to the liquid per unit weight, obtained by dividing the increase in pressure by the specific weight. This specific energy is determined by the Bernoulli equation. Height may be defined as the height to which 1 kg of discharged liquid can be lifted by the energy supplied by a pump. Therefore, it does not depend on the specific weight or density of liquid to be pumped;

③ Power $N(kgf \cdot m/s)$—the energy consumed by a pump per unit time for supplying liquid energy in the form of pressure. Power is equal to the product of specific energy, H, and the mass flowrate, γQ: $N = \gamma QH = \rho gQH$;

Effect power, N_e, is larger than N due to energy losses in a pump, its relative value is evaluated by the pump efficiency.

④ Overall efficiency—the ratio of useful hydraulic work performed to the actual work input. This parameter characterizes the perfection of design and performance of a pump.

Selected from "Reading Material for Process Equipment", Yin Xuwei, Chemical Engineering Press, 1997.

New Words and Expressions

1. refrigeration [rifridʒə'reiʃən] *n*. 冷藏，制冷，冷却
2. sodium ['səudjəm] *n*. [化] 钠
3. centrifugal [sen'trifjugəl] *a*. 离心的
4. diaphragm ['daiəfræm] *n*. 隔膜，(电话等) 振动膜
5. plunger ['plʌndʒə] *n*. 柱塞
6. leaktight chamber 防漏腔
7. hydrocarbon ['haidrəu'kɑːbən] *n*. 烃，碳氢化合物
8. screw pump 螺杆泵，螺旋泵
9. offset ['ɔfset] *n*. 偏移量，抵销，弥补
10. crescent-shaped ['kresntʃeipt] *a*. 新月形的，逐渐增加的
11. elastomer [iˈlæstəumə] *n*. 弹性体，人造橡胶

12. self-priming [selfpraimiŋ] *a*. 自吸的
13. back-and-forth [bæk ænd fɔːθ] *ad*. 前后（运动），来回（运动）
14. capacity [kəˈpæsiti] *n*. 容量，生产量，生产力，能力
15. head [hed] *n*. 扬程
16. specific weight 相对密度

Unit 14 • Quality Assurance and Control (1)

Product quality is of paramount importance in manufacturing. If quality is allowed to deteriorate, then a manufacturer will soon find sales dropping off followed by a possible business failure. Customers expect quality in the products they buy, and if a manufacturer expects to establish and maintain a name in the business, quality control and assurance functions must be established and maintained before, throughout, and after the production process①. Generally speaking, quality assurance encompasses all activities aimed at maintaining quality, including quality control. Quality assurance can be divided into three major areas. These include the following:

① Source and receiving inspection before manufacturing;

② In-process quality control during manufacturing;

③ Quality assurance after manufacturing.

Quality control after manufacture includes warranties and product service extended to the users of the product.

Source and Receiving Inspection before Manufacturing

Quality assurance often begins long before any actual manufacturing takes place. This may be done through source inspections conducted at the plants that supply materials, discrete parts, or subassemblies to manufacturer. The manufacturer's source inspector travels to the supplier factory and inspects raw material or premanufactured parts and assemblies. Source inspections present an opportunity for the manufacturer to sort out and reject raw materials or parts before they are shipped to the manufacturer's production facility.

The responsibility of the source inspector is to check materials and parts against design specifications and to reject the item if specifications are not met. Source inspections may include many of the same inspections that will be used during production. Included in these are:

① Visual inspections;

② Metallurgical testing;

③ Dimensional inspection;

④ Destructive and nondestructive inspection;

⑤ Performance inspection.

Visual Inspections

Visual inspections examine a product or material for such specifications as color, texture, surface finish, or overall appearance of an assembly to determine if there are any obvious deletions of major parts or hardware.

Metallurgical Testing

Metallurgical testing is often an important part of source inspection, especially if the

primary raw material for manufacturing is stock metal such as bar stock or structural materials. Metals testing can involve all the major types of inspections including visual, chemical, spectrographic, and mechanical, which include hardness, tensile, shear, compression, and spectrographic analysis for alloy content. Metallurgical testing can be either destructive or nondestructive.

Dimensional Inspection

Few areas of quality control are as important in manufactured products as dimensional requirements. Dimensions are as important in source inspecting as they are in the manufacturing process. This is especially critical if the source supplies parts for an assembly. Dimensions are inspected at the source factory using standard measuring tools plus special fit, form, and function gages that may be required. Meeting dimensional specifications is critical to interchangeability of manufactured parts and to the successful assembly of many parts into complex assemblies such as autos, ships, aircraft, and other multipart products②.

Destructive and Nondestructive Inspection

In some cases it may be necessary for the source inspections to call for destructive or nondestructive tests on raw materials or parts and assemblies. This is particularly true when large amounts of stock raw materials are involved. For example it may be necessary to inspect castings for flaws by radiographic, magnetic particle, or dye penetrant techniques before they are shipped to the manufacturer for final machining. Specifications calling for burn-in time for electronics or endurance run tests for mechanical components are further examples of nondestructive tests.

It is sometimes necessary to test material and parts to destruction, but because of the costs and time involved destructive testing is avoided whenever possible. Examples include pressure tests to determine if safety factors are adequate in the design. Destructive tests are probably more frequent in the testing of prototype designs than in routine inspection of raw material or parts. Once design specifications are known to be met in regard to the strength of materials, it is often not necessary to test further parts to destruction unless they are genuinely suspect.

Performance Inspection

Performance inspections involve checking the function of assemblies, especially those of complex mechanical systems, prior to installation in other products. Examples include electronic equipment subcomponents, aircraft and auto engines, pumps, valves, and other mechanical systems requiring performance evaluation prior to their shipment and final installation③.

Selected from "Modern Materials and Manufacturing Process", R. Gregg Bruce, Mileta M. Tomovic, John E. Neely and Richard R. Kibbe, Prentice-Hall, Inc. 1998.

New Words and Expressions

1. **paramount** [ˈpærəmaunt] *a*. **最重要的，头等的，最高的**

2. deteriorate [di'tiəriəreit] *v.* (使) 恶化，退化
3. assurance [ə'ʃuərəns] *n.* 保证，担保
4. encompass [in'kʌmpəs] *v.* 包含，包括
5. inspection [in'spekʃən] *n.* 检验，检查
6. in-process *a.* 在制（加工，处理）过程中的
7. warranty ['wɔrənti] *n.* 担保书，保证书
8. premanufacture [primænju'fæktʃə] *n.* 预制造
9. visual inspection 肉眼检查，目测检验
10. metallurgical [ˌmetə'ləːdʒikəl] *a.* 冶金的，冶金学的
11. destructive [di'strʌktiv] *a.* 毁灭性的，破坏的，有害的
12. nondestructive inspection 无损检测
13. texture ['tekstʃə] *n.* 纹理，[织物的] 密度，[材料的] 结构
14. spectrographic ['spektrəugrɑːfik] *a.* 摄谱仪的，光谱的
15. gage [geidʒ] *n.* 标准度量，计量器，量具
16. endurance [in'djurəns] *n.* 持久性，耐久性

Notes

① 参考译文：顾客们希望他们所购买的产品质量好，而若制造商期望树立与维持商业名声，则遍及生产过程的前后必须建立与维护质量控制和质量保证。

② 参考译文：符合尺寸规格对所制造部件的互换性和对多部件成功组装成复杂的装置如：汽车、轮船、飞机和其他多部件产品都是极其重要的。

③ 参考译文：例如电子设备的组件、飞机和汽车的发动机、泵、阀及其他机械系统，在发货和最终安装前都需要进行性能检测评估。

Exercises

1. After reading the text above, write a summary of it.
2. Answer the following questions according to the text.
   ① What are the three major areas of quality assurance?
   ② What items does the source inspector look for?
   ③ What is the difference between destructive and nondestructive testing?
   ④ What is a performance inspection?
3. Translate the 6th paragraph of the text into Chinese.
4. Put the following into Chinese by reference to the text.

   raw material　　compression　　casting
   radiographic　　prototype　　visual inspection

5. Put the following into English.

   质量控制　　无损检测　　性能　　冶金
   硬度　　计量器

6. Translate the following sentences into English.
   ① 质量保证分为三部分：制造前原材料检验、制造中质量控制及制造后的质量保证。
   ② 性能检测包括检测组件的功能，尤其对于复杂机械系统的组件，要在组装成产品之前进行性能测试。

Reading Material 14

Quality Assurance and Control (2)

Quality Control during Manufacturing

Quality control during manufacturing involves many of the same checks as in source inspection. A great deal of time and effort is invested in in-process quality control by the manufacturing industry. Major considerations for in-process quality control include the following:

① Receiving inspection;

② First piece inspection;

③ In-process inspection;

④ One hundred percent inspection;

⑤ Final inspection.

Receiving Inspection

In receiving inspection, parts, material, and assemblies are checked against design specifications as they arrive at the manufacturer's production or assembly facility. If the items received are premanufactured parts requiring further assembly or processing, receiving inspection will inspect for correct dimensions and any other specifications required by the design.

First Piece Inspection

Any time a new part or product is made for the first time or when a production machine is set up for repeat production after having been used for another job, a first piece inspection is required. When the first item completes the manufacturing process it is inspected, and if it conforms to specifications further production is allowed to continue. If the first piece is not acceptable, the machine or in some cases the entire production line must be stopped and reset to correct any problems. First piece inspection is necessary before production can begin.

In-process Inspection

In-process inspection must be applied continuously during production. Each production station must have the necessary tools to accomplish the required inspection operations. Past practice of employing roving quality inspectors to randomly inspect parts throughout the factory has largely been replaced by in-process inspection performed by the machine operator or assembler at his or her workstation. In this manner the worker, often aided by statistical process control (SPC) measures, is made responsible for product quality.

One Hundred Percent Inspection

One hundred percent inspection refers to the inspection of every part or assembly produced. Though this might appear to be a way to solve all quality control problems, it is often impractical and altogether too expensive to accomplish in routine manufacturing. This is especially true in high-volume production where it would not be feasible to inspect each and every item produced. Also, one hundred percent inspection does not always produce perfect

quality. Inspection operations can be fallible and fail to detect defective products, especially in the case of highly repetitive manual inspection.

Regardless of its faults, one hundred percent inspection is often used for critical components. Critical aircraft parts may be one hundred percent inspected since the possible acceptance of an out-of-specification part could have serious consequences. Another place where such inspection may be applied is where nonconforming parts begin to show up continually in the production process, and the problem must be tracked down and immediately corrected. One hundred percent inspection may be necessary until production quality is restored, when normal in-process inspection may resume.

Final Inspection

Final inspection presents another opportunity to sort out parts or assemblies that do not meet specifications. Nonconforming parts are rejected to be scrapped or recycled for rework if possible. It is naturally desirable to avoid scrapping as many parts as possible since they often represent a large production expense.

Quality Assurance after Manufacturing

Quality control and assurance do not end at the manufacturer's shipping dock. They continue well after manufacture in the form of warranties providing factory or factory-authorized service, spare parts, field service, service contracts, reliability studies, failure reports and recommendations, and repair and reconditioning product support.

Purchasers of almost any product expect and are entitled to a warranty or guarantee that will be in force for a period of time after the product arrives in the hands of the end user. Warranties vary in length ranging from a few months to lifetime full replacement. Some warranties cover full repair and replacement for a short period of time and cover only specific components for an extended time. Extended warranties may be purchased at extra cost in many cases. Related to this are service contracts for which the customer pays a flat extra cost, sometimes a percentage of the purchase price, in exchange for repair service for the duration of the contract.

Critical plant equipment that fails in service must often be restored to functional capability as quickly as possible, and efficient service can be critical to the continuing function of plants and businesses. Many manufacturers provide factory service on their products. This capability may be extended over wide geographical areas by using factory-authorized service through other businesses. Spare parts can often be purchased from either the original maker or a factoryauthorized service center.

Many manufacturers, especially those of high-technology products, are concerned with reliability and failure rates of their products in service. Product failure and reliability analysis is useful in determining where improvements or correction of deficiencies in the production or design need to be made. Reliability analysis can greatly improve a manufacturer's product quality. The compilation of reliability statistics can be used effectively in product advertising, thus enhancing the manufacturer's standing in the business.

Product Testing

Product testing is vital to successful manufacturing, and manufacturing industries often place a large emphasis on product testing. Testing may take place at the prototype stage before an actual product is manufactured for sale. Product testing also may be carried out in controlled markets where a few preproduction models are placed in selected hands for the purpose of testing and evaluation. In the case of aircraft, for example, a license to manufacture will not be granted until design and safety specifications have been met. The manufacturer must assume the costs of manufacturing and testing preproduction prototypes.

Endurance testing of automobiles is another significant product testing activity. Production models are test driven through many different environments to simulate what the vehicles will encounter in actual service. By continuous testing, several years of typical service of the vehicle can be compressed into a few months. The data collected can be invaluable in determining design problems and in predicting the reliability and effective service life of the product.

Selected from "Modern Materials and Manufacturing Process (second edition)", R. Gregg Bruce, Mileta M. Tomovic, John E. Neely and Richard R. Kibbe, Published by Prentice-Hall, Inc. 1998.

New Words and Expressions

1. conform to 与相符［一致］，符合，遵守
2. reset [ˈriːset] *vt.*; *n.* 重新调整，重新安装
3. roving [ˈrəuviŋ] *a.* 流动的
4. impractical [imˈpræktikəl] *a.* 不（切）实际的，不现实的
5. fallible [ˈfæləbl] *a.* 易错的，可能犯错的
6. recondition [ˈriːkənˈdiʃən] *vt.* 修理［复，整，补］，检［翻，整］修
7. repetitive [riˈpetitiv] *a.* 重复的，反复性的
8. invaluable [inˈvæljuəbl] *a.* 无价的，价值无法衡量的

PART Ⅱ

FUNDAMENTALS OF AUTOMATION

Unit 15 • Control System Fundamentals

Numerically controlled machines often weigh up to 100 tons and yet are required to position a cutting tool with an accuracy of the order of 0.002 mm. The control system must move the tool at feedrates as high as 8 cm/sec while encountering loads which may vary dramatically on a given path. The NC machine must have dynamic response characteristics that enable it to follow intricate contours with a minimum of path error. Clearly, these requirements dictate a control system that is matched to the mechanical characteristics of the machine it drives.

A control system is a combination of devices that regulate an operation by administering the flow of energy and other resources to and from that operation. Essentially, a control system is made up of interrelated subsystems that perform tasks which in orthodox machining processes are managed by an intelligent human operator.

The control systems for NC machines therefore serve to replace the human machine operator and significantly improve upon even the best human performance. A direct analogy can be made between numerical controls and the human operators they replace. Both:

① Sense the current status of the machine;

② Make logical decisions which are required to accomplish a task;

③ Communicate decisions to the machine by actuating proper mechanical devices;

④ Have the ability to store information: instructions, data, and the results of logical decisions.

In summary, a machine control system is a combination of electronic circuitry, sensing devices, and mechanical components which guide the cutting tool along a predefined path.

In the sections that follow, the reader is introduced to the theory of machine control.

We have attributed four major characteristics to any control system: the ability to sense data; make logical decisions; communicate and actuate; and store information in a memory①. If a system performs these functions without the aid of a human operator, by definition it automatically controls a given task. Let us consider one such simple operation.

Suppose that an electric motor is used to a freight conveyor between two floors of a building. To operate smoothly the conveyor must maintain a constant speed as additional loads are added from loading shoots. We can assume that the speed of rotation of an electric

motor is directly proportional to the electric current supplied to it. Hence as the load increases on the conveyor, the torque requirement of the motor is increased and additional power must be supplied to maintain constant speed. This is usually achieved by increasing the supply voltage.

If a motor speed indicator is within sight of a human operator, he can control the conveyor speed by increasing or decreasing the voltage supply manually. Such a system is illustrated schematically in Fig. 15. 1.

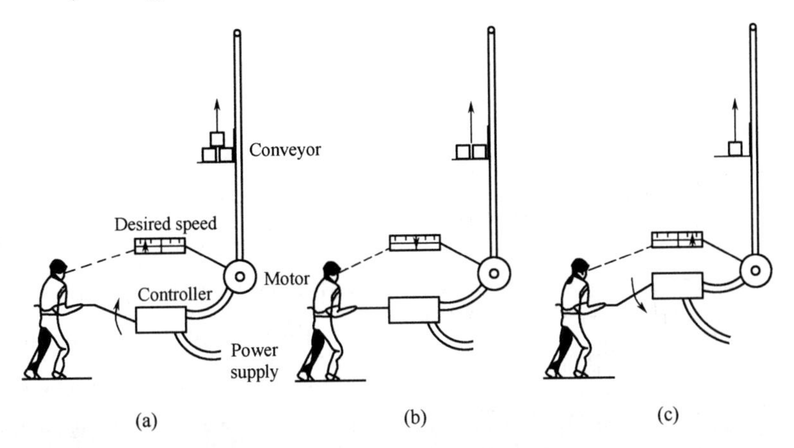

Fig. 15. 1 A manually controlled conveyor

As indicated by Fig. 15. 1(a), the operator first senses that the speed is below the desired value. Logically, he determines that an increase in power will increase motor speed. He communicates a decision, by actuating the power supply control lever, to increase voltage until the desired speed reading is obtained. In Fig. 15. 1(b) the desired speed is read; hence, the operator takes no action. If the load is decreased so that a speed greater than the desired value is registered in Fig. 15. 1(c), the operator will decrease voltage.

The process described above is illustrated schematically in Fig. 15. 2. The system indicated in the figure is not automatic. However, if the human operator block were replaced with a device that performed the functions shown, an automatic control system would be realized.

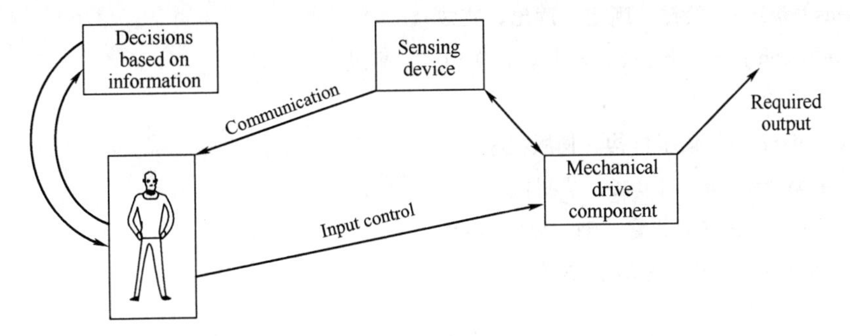

Fig. 15. 2 Schematic of information flow for conveyor system

Automatic control systems are sometimes called servomechanisms. A servomechanism may be a complex array of electromechanical components or it may be a simple mechanical device. Regardless of the complexity of the servomechanism, it always exhibits the features illustrated in Fig. 15. 2.

Consider a simple mechanical device that performs the automatic control function for our conveyor system (Fig. 15. 3). Such a device, called a Watt speed governor[②], can automatically control the speed of an electric motor.

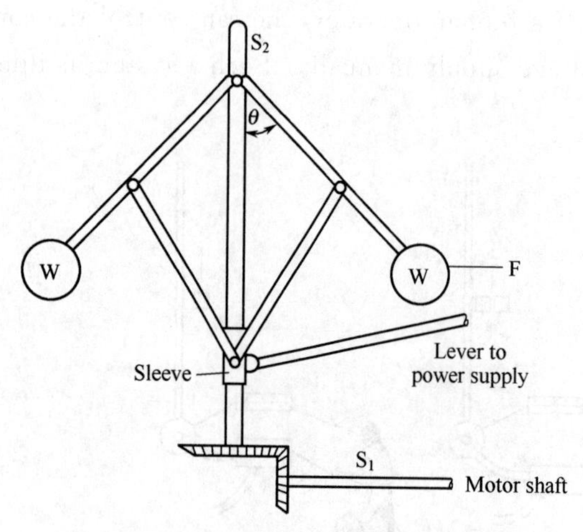

Fig. 15. 3 The Watt mechanism-an early example of an automatic control unit

As shown in Fig. 15. 3, the Watt speed governor consists of two weighted arms connected to a shaft, S_2, and a sleeve. The motor shaft, S_1, is coupled to the governor shaft, S_2, so that the rotational velocity of S_2 is proportional to S_1. As motor speed increases, centrifugal force moves the weights, W, outward, causing the weighted arms to rotate through some angle, θ. The weights, angle, and sleeve position are chosen so that the proper speed will cause the voltage supply lever to be positioned at a neutral position. Variations in speed will cause the appropriate movement of the sleeve and lever.

Selected from "Numerical Control and Computer-Aided Manufacturing", Roger S. Pressman and John E. Williams, John Wiley & Sons, Inc., 1977.

New Words and Expressions

1. and yet 可是，(然) 而，但
2. feedrate [ˈfiːdreit] *n*. 馈送率，进料速度，进给速度
3. dramatically [drəˈmætikəli] *ad*. 显著地，引人注目地
4. orthodox [ˈɔːθədɔks] *a*. 传统的，正统的，惯常的，普通的
5. improve upon (对……加以) 改进 (良)，作出比……更好的东西
6. sense [sens] *vt*.; *n*. 检测，断定，读出，感知
7. actuate [ˈæktjueit] *vt*. 开 [驱，起] 动，使动作，操纵
8. attribute *M* to *N* 把 *M* 赋予 *N*
9. servo mechanism *n*. 伺服机构，伺服机械
10. governor [ˈɡʌvənə] *n*. 调节器，节制器
11. sleeve [sliːv] *n*. 套 (筒、管、轴、环)，轴套
12. couple *M* to [with] *N* 使 *M* 同 *N* 结 [配、耦] 合

Notes

① 参考译文：我们认为任何控制系统具有四个主要特征：检测数据的能力，作出合乎逻辑的决策，传递决策并操纵机器，把信息储存在储存器内。

② 参考译文：研究一种实现我们的传送带自动控制功能的简单机械装置，这种装置称

作瓦特调速器。

Exercises

1. After reading the text above, summarize the main ideas of it in oral English.
2. Answer the following questions according to the text.
   ① What do the control systems of NC machines serve to?
   ② What is a machine control system?
   ③ What are the four major characteristics of any control system?
   ④ Describe the mechanism of the Watt speed governor.
3. Translate the 2nd and 3rd paragraphs into Chinese.
4. Put the following into Chinese by reference to the text.
   contour　orthodox　sense　conveyor　servomechanism　governor
5. Put the following into English.
   进料速度　动态响应　储存信息　与……成正比
   速度表　转矩　杠杆
6. Translate the following sentences into English.
   ① 检测机器的现状是控制系统的特征之一。
   ② 机器控制系统是电路系统、检测装置以及机械零件的结合体。

Reading Material 15

Applications of Automatic Control

Feedback control systems are to be found in almost every aspect of our daily environment. In the home, the refrigerator utilizes a temperature-control system. The desired temperature is set and a thermostat measures the actual temperature and the error. A compressor motor is utilized for power amplification. Other applications of control in the home are the hot-water heater, the central heating system, and the oven, which all work on a similar principle.

In industry, the term automation is very common. Modern industrial plants utilize temperature controls, pressure controls, speed controls, position controls, etc. The chemical process control field is an area where automation has played an important role. Here, the control engineer is interested in controlling temperature, pressure, humidity, thickness, volume, quality, and many other variables. Areas of additional interest include automatic warehousing, inventory control and automation of farming.

Modern control concepts are being utilized in an ever increasing degree to help solve various problems. In the transportation field, for example, automatic control systems have been devised to regulate automobile traffic and control high-speed train systems. A very widely acclaimed high-speed rail transportation system is in operation in Japan. This latest innovation to the system of the Japanese National Railways is the Tokyo-to-Osaka super-express train. The high-speed railroad link between Tokyo and Osaka is commonly called the

Tokaido line. It travels over a 320-mile route in 3 hours and 10 minutes. The train can travel at 130 miles per hour over most of the route. The system utilizes a control computer to control the trains in an optimum manner. Fig. 15. 4 illustrates the general position control concepts of such a high-speed automated train system. Observe from the diagram that it contains a position-measuring loop and a velocity-measuring loop. Position can be measured from the rotation of the train's wheels. Speed can be measured by using velocity-sensing devices such as tachometers. The control computer system monitors the positions and speeds of all trains in the system and issues control signals via a high-speed communication system.

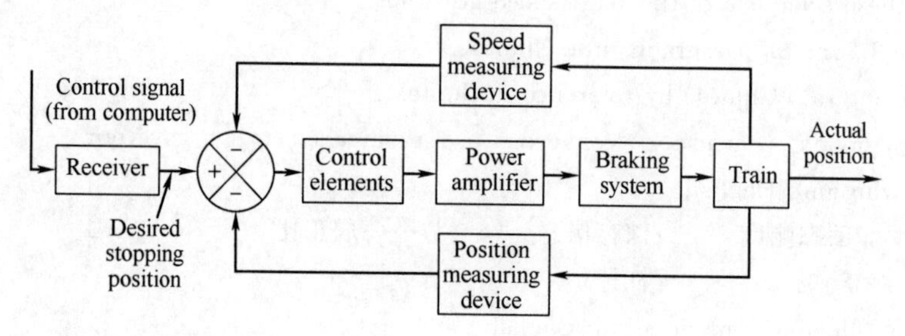

Fig. 15. 4 Automatic position control system

Automatic control systems have been applied to a large degree by the aerospace industry. Modern high-speed aircraft, such as the F-4 long-range all-weather interceptor and attack bomber, are controlled almost entirely automatically during their missions.

The Polaris Poseidon Fleet Ballistic Missile (FBM) Weapon System is a good example of the application of automatic control to a complex weapon system. Let us go through the process of launching and guiding this missile in order to illustrate the multitude of automatic control systems involved in the operation. Polaris missiles are launched from the submarine by means of an air or a gas/steam generator ejection system. The missile is propelled from the launch tube, through the water and to the surface. At this point, a control system automatically ignites the missile's first-stage rocket motor and sends the missile on its mission. Two positions must be known very accurately for successfully controlling a missile: that of the target and launcher. In addition, the initial velocity of the launcher must be known very accurately. Since the position and velocity of the ship (launcher) is continuously changing in the FBM system, great emphasis must be placed on the design of the navigation system. The FBM navigation system, managed by the Sperry Systems Management Division, Sperry Rand Corporation, Great Neck, New York, utilizes several methods that complement each other in order to provide a very high order of accuracy in determining the ship's position and velocity. The heart of the system is the Ship's Inertial Navigation System (SINS), a complex system of gyroscopes, accelerometers, and computers, which relate movement and speed of the ship in all directions to true north in order to give a continuous report of ship position and velocity. The missile's guidance package consists of an inertial platform and a digital computer. The inertial platform is a gyro-stabilized set of three accelerometers. Once launch has occurred, the missile computer is in complete control of the missile. During flight, missile accelerations are measured by the inertial platform and integrated into velocities which are continuously fed to the computer. The computer continuously compares the attained velocity information with that reference velocity which will permit the payload to continue on to the target on a ballistic trajectory. When this desired velocity is attained, the computer automatically issues a signal which commands separation from the second stage motor and the payload continues on a ballistic trajectory to the target.

Automatic control theory has also been applied for modeling the feedback processes of our economic system in order to understand it better. Our economic system contains many feedback systems and regulatory agencies.

Feedback control systems have a very bright and unlimited potential. This important field is one in which the engineer should become knowledgeable and proficient for solving problems found in this modern technological age.

Selected from "Modern Control System Theory and Application", Stanley M. Shinners, Addison-Wesley Publishing Company, Inc., 1978.

Selected from "Modern Control System Theory and Design", Stanley M. Shinners, John Wiley & Sons, Inc., 1992.

New Words and Expressions

1. thermostat [ˈθəːməstæt] *n*. 恒温器，温度自动调节器
2. amplification [ˌæmplifiˈkeiʃən] *n*. 放大，扩大
3. interceptor [ˌintəˈseptə] *n*. 拦截战斗机
4. polaris/poseidon fleet Ballistic Missile (FBM) 北极星式/海神式舰载弹道导弹
5. ejection [iˈdʒekʃən] *n*. 发射，放射，弹射
6. propel [prəˈpel] *vt*. 推进
7. ignite [igˈnait] *vt*. 点火，点燃
8. gyroscope [ˈdʒairəskəup] *n*. 陀螺仪，回转仪
9. accelerometer [ækˌseləˈrɔmitə] *n*. 加速度计
10. gyro-stabilized *a*. 陀螺稳定的
11. trajectory [træˈdʒiktəri] *n*. 轨道，弹道，轨迹
12. proficient [prəˈfiʃənt] *a*.；*n*. 精通的；熟练的；高手，专家

Unit 16 • Open-Loop and Closed-Loop Control

Open-Loop Control Systems (Nonfeedback Systems)

The word automatic implies that there is a certain amount of sophistication in the control system. By automatic, it generally means that the system is usually capable of adapting to a variety of operating conditions and is able to respond to a class of inputs satisfactorily. However, not any type of control system has the automatic feature. Usually, the automatic feature is achieved by feeding the output variable back and comparing it with the command signal. When a system does not have the feedback structure, it is called an open-loop system, which is the simplest and most economical type of control system. Unfortunately, open-loop control systems lack accuracy and versatility and can be used in none but the simplest types of applications.

Consider, for example, control of the furnace for home heating. Let us assume that the furnace is equipped only with a timing device, which controls the on and off periods of the furnace. To regulate the temperature to the proper level, the human operator must estimate the amount of time required for the furnace to stay on and then set the timer accordingly. When the preset time is up, the furnace is turned off. However, it is quite likely that the house temperature is either above or below the desired value, owing to inaccuracy in the estimate. Without further deliberation, it is quite apparent that this type of control is inaccurate and unreliable. One reason for the inaccuracy lies in the fact that one may not know the exact characteristics of the furnace. The other factor is that one has no control over the outdoor temperature, which has a definite bearing on the indoor temperature. This also points to an important disadvantage of the performance of an open-loop control system, in that the system is not capable of adapting to variations in environmental conditions or to external disturbances. In the case of the furnace control, perhaps an experienced person can provide control for a certain desired temperature in the house; but if the doors or windows are opened or closed intermittently during the operating period, the final temperature inside the house will not be accurately regulated by the open-loop control.

An electric washing machine is another typical example of an open-loop system, because the amount of wash time is entirely determined by the judgment and estimation of the human operator. A true automatic electric washing machine should have the means of checking the cleanliness of the clothes continuously and turn itself off when the desired degree of cleanliness is reached.

The elements of an open-loop control system can usually be divided into two parts: the controller and the controlled process, as shown in Fig. 16. 1. An input signal or command r is applied to the controller, whose output acts as the actuating signal e; the actuating signal then controls the controlled process so that the controlled variable c will perform according to prescribed standards. In simple cases, the controller can be an amplifier, mechanical linkage, filter, or other control element, depending on the nature of the system. In more sophisticated cases, the controller can be a computer such as a microprocessor. Because of the simplicity and economy of open-loop control systems, we find this type of system in many noncritical applications.

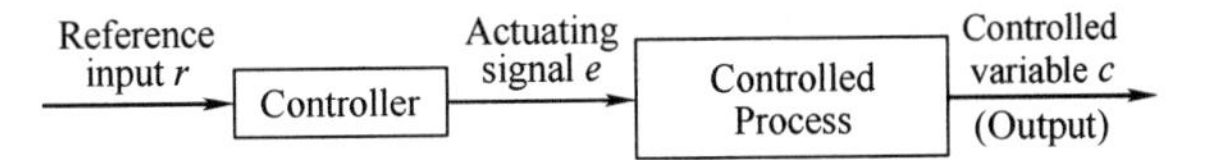

Fig. 16. 1 Elements of an open-loop control system

Closed-Loop Control Systems (Feedback Control Systems)

What is missing in the open-loop control system for more accurate and more adaptable control is a link or feedback from the output to the input of the system. In order to obtain more accurate control, the controlled signal $c(t)$ must be fed back and compared with the reference input, and an actuating signal proportional to the difference of the output and the input must be sent through the system to correct the error①. A system with one or more feedback paths like that just described is called a closed-loop system. Human beings are probably the most complex and sophisticated feedback control system in existence. A human being may be considered to be a control system with many inputs and outputs, capable of carrying out highly complex operations.

To illustrate the human being as a feedback control system, let us consider that the objective is to reach for an object on a desk. As one is reaching for the object, the brain sends out a signal to the arm to perform the task. The eyes serve as a sensing device which feeds back continuously the position of the hand. The distance between the hand and the object is the error, which is eventually brought to zero as the hand reaches the object. This is a typical example of closed-loop control. However, if one is told to reach for the object and then is blindfolded, one can only reach toward the object by estimating its exact position. It is quite possible that the object may be missed by a wide margin. With the eyes blindfolded, the feedback path is broken, and the human is operating as an open-loop system. The example of the reaching of an object by a human being is described by the block diagram shown in Fig. 16. 2.

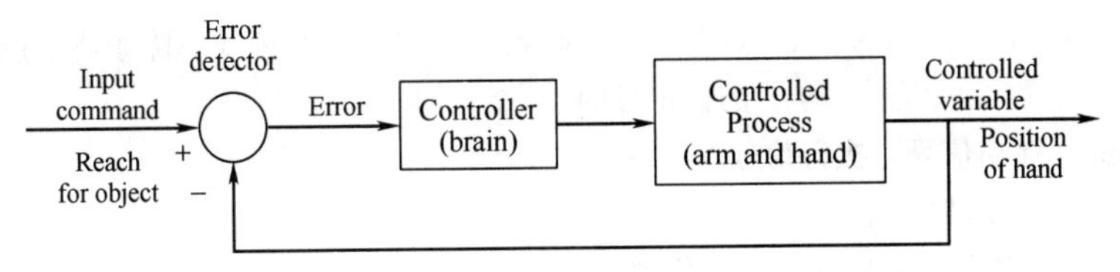

Fig. 16. 2 Block diagram of a human being as a closed-loop control system

As another illustrative example of a closed-loop control system, Fig. 16. 3 shows the block diagram of the rudder control system of a ship. In this case the objective of control is the position of the rudder, and the reference input is applied through the steering wheel. The error between the relative positions of the steering wheel and the rudder is the signal, which

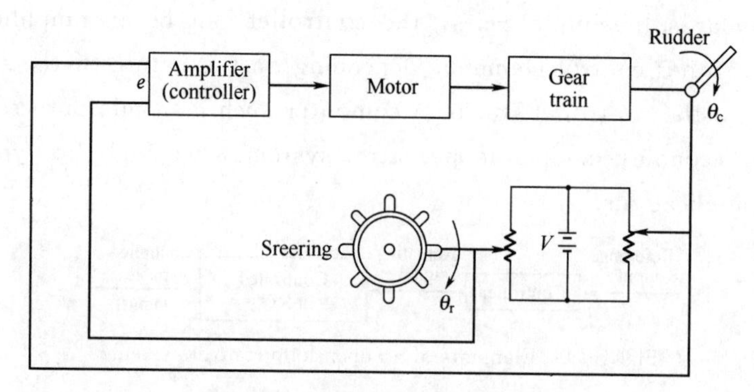

Fig. 16. 3 Rudder control system

actuates the controller and the motor. When the rudder is finally aligned with the desired reference direction, the output of the error sensor is zero.

The basic elements and the block diagram of a closed-loop control system are shown in Fig. 16. 4. In general, the configuration of a feedback control system may not be constrained to that of Fig. 16. 4. In complex systems there may be a multitude of feedback loops and element blocks.

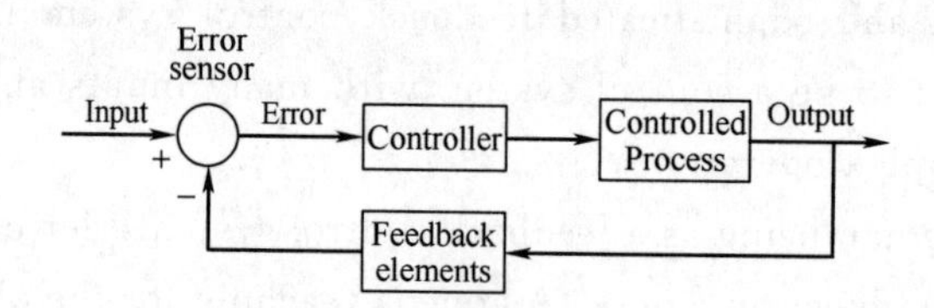

Fig. 16. 4 Basic elements of a feedback control system

Selected from "Automatic Control System", Benjamin C. Kuo, Prentice-Hall, Inc. , 1975 and "Automatic Control Systems", Benjamin C. Kuo, Prentice-Hall, Inc. , 1995.

New Words and Expressions

1. sophistication [səˌfistiˈkeiʃən] *n.* 复杂化，完善（化），采用先进技术
2. versatility [ˌvəːsəˈtiləti] *n.* 多面性，多功能性，多方面适应性，多样性
3. desired value 预期值
4. deliberation [diˌlibəˈreiʃən] *n.* 考虑，熟虑，熟思
5. intermittent [ˌintə(ː)ˈmitənt] *a.* 间歇的，断断续续的
6. blindfold [ˈblaindʃəuld] *vt.*；*n.*；*a.* 蒙住……的眼睛，遮住……的视线；障眼物，遮眼物；看不清的，盲目的
7. actuating signal 促动信号，执行信号
8. rudder [ˈrʌdə] *n.* 方向舵，舵
9. steering wheel *n.* 舵轮，转向［方向，驾驶］盘
10. align [əˈlain] *v.* 对中，（使，排）成一直线，调整

Notes

① 参考译文：为了获得更准确控制，必须使受控信号 $c(t)$ 反馈并与参考输入相比较，而正比于输出与输入差的驱动信号则必须送经系统以校正误差。

Exercises

1. After reading the text above, write a summary of it.
2. Answer the following questions according to the text.
   ① What does the word automatic mean?
   ② What is the difference between the open-loop control and closed-loop control system?
   ③ Cite a disadvantage of the performance of an open-loop control system.
   ④ How does a closed-loop control system work?
3. Translate the 1st paragraph of the text into Chinese.
4. Put the following into Chinese by reference to the text.
   open-loop control　　closed-loop control　　feedback　　output variable
   actuating signal　　desired value
5. Put the following into English.
   框图，原理图　　看不清的，盲目的　　干扰，扰动　　误差检测器
   放大器　　反馈控制系统

Reading Material 16

The Modes of Control Action

The operational response of a controller is often described as its mode of control. Several different types of control are available. In some cases, only a single mode of control is needed to accomplish an operation. This is described as a pure control operation. On-off, proportional, integral, and derivative are examples of pure control. More sophisticated control is achieved by combining two or more pure modes of operation. This is described as a composite mode. Proportional plus integral, proportional plus derivative, and proportional plus integral plus derivative are examples of composite control modes. The specific mode of operation utilized by an instrument is determined by the control procedure of the application.

On-off operation

An on-off or two-state controller is the simplest of all process control operations. The actuator or final control element driven by the output of the controller is automatically switched on or off. It does not have any intermediate level of operation. Control of this type is popular and inexpensive to accomplish.

The home heating system in Fig. 16. 5 is a common example of on-off control. The thermostat of this system serves as the controller. Should the interior temperature of the building drop below the setpoint value of the thermostat, the furnace will turn on. In a gas-furnace heating system, a solenoid gas valve serves as the final control element. It is energized by the thermostat and turns on the source of heat. Heat from this assembly causes the interior temperature of the home to rise. When the temperature rises above the setpoint value, the thermostat tells the furnace to stop producing heat. The solenoid gas valve closes and turns off the heat source. The heat source continues to be off until the temperature again drops below

the setpoint of the thermostat. When this occurs, the thermostat again turns on the gas valves. This on-off cycling action continues according to the demands of the system as long as the system is operational. This response is typical of an on-off controller.

The operation of an on-off controller is determined by the position adjustment of its setpoint value. As a rule, the output continually cycles or oscillates above or below the setpoint value. Fig. 16. 6 shows the cycling response of an on-off temperature controller. Note the location of the setpoint and how the temperature of the system rises and falls above this value. The final control element of this system is the gas solenoid valve. The source of gas is turned on and off with respect to time. When gas is applied, it produces heat. Turning off the gas stops the process. In systems where continuous on-off cycling occurs, there can be some damage to the final control element after prolonged operational periods. On-off control is generally used in systems where precise control is not necessary. These systems must have a large capacity that changes slowly in order to be effective.

Assume that the setpoint of the thermostat is adjusted to 70 ℉. If the temperature is less than 70 ℉ by a tiny amount, the gas valve will be fully on, or 100 percent open. This causes an immediate increase in temperature.

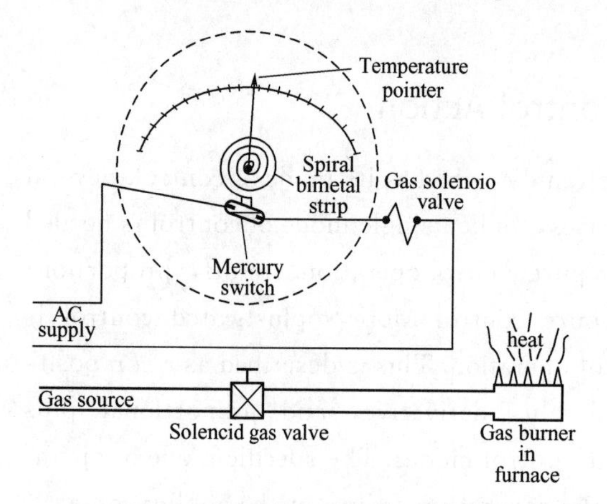

Fig. 16. 5 Thermostat-controlled gas heating system.

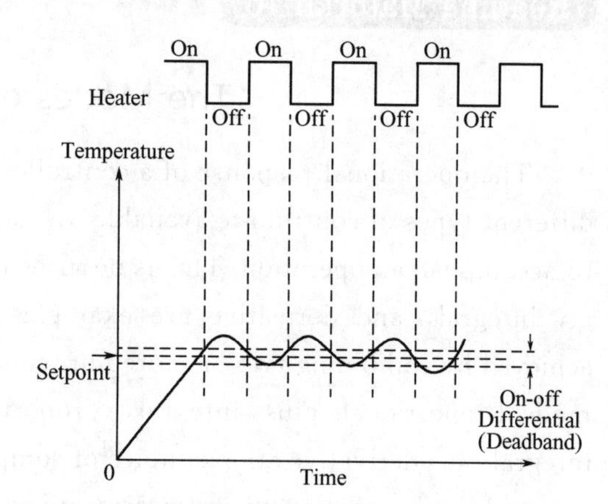

Fig. 16. 6 On-off temperature control action

When the temperature rises above 70 ℉ by a tiny amount, it causes the gas valve to be fully off, or closed. This means that the gas valve of our ideal heating system continually cycles on and off, which can cause the gas valve to be damaged after a prolonged operational period.

Proportional control

In on-off control, the final control element was either on or off. If the control element were a valve, it would have been fully open or closed. There is no intermediate adjustment of the valve. In proportional control, the final control element can be adjusted to any value between fully open and fully closed. Its value is determined by a ratio of the setpoint input and the actual process value of the system. In a valve-controlled system, operation is arranged so that the valve is normally adjusted to some percentage of its operating range.

Refer to the water temperature control system in Fig. 16. 7. In this system, cool water enters the tank at a constant temperature. It exits the tank at a higher but rather constant temperature. Steam applied to the heating coil increases the temperature of the water to some predetermined value. This value is determined by the setpoint adjustment. The controller senses the temperature of the water in the tank and compares the actual temperature of the water with the predetermined setpoint value. The motor-driven steam valve is actuated by the proportional controller output. Valve opening is determined by the difference in temperature between actual and setpoint values.

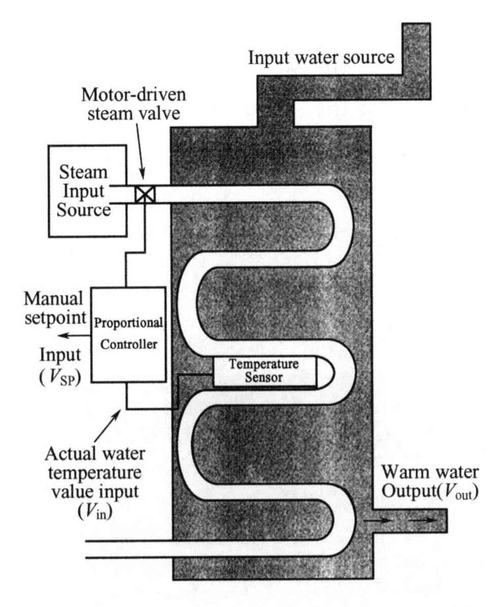

Fig. 16. 7 Water temperature control system

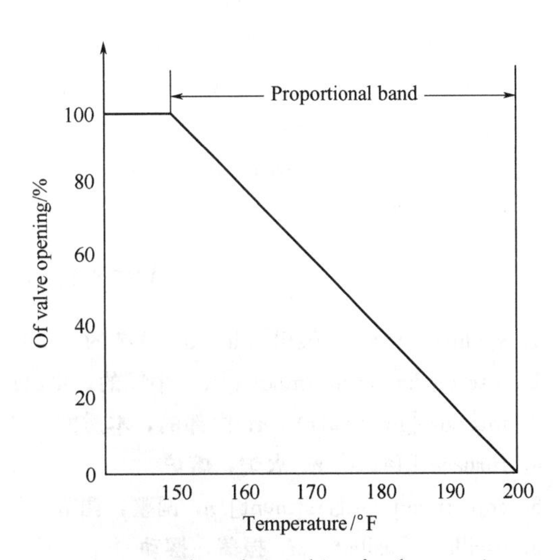

Fig. 16. 8 The relationship of valve opening percentage to water temperature

The graph in Fig. 16. 8 shows a relationship between valve opening percentage and water temperature. The valve can be 100 percent open or fully closed. The water temperature range is 150 ℉ to 200 ℉. The difference between temperature values for the valve to be fully open or closed is called the proportional band of control. The proportional band of this system is 200～150 ℉, or 50 ℉.

The proportional band of a controller is generally expressed as a percentage of the full-scale span of operation. As a rule, the setpoint adjustment range of the controller usually identifies the operational span of the proportional band. The operational span of the setpoint is usually greater than the value of the proportional band. For our water-heating controller, the proportional band was 50 ℉. The setpoint adjustment span could be from 250 ℉ to 100 ℉, or 150 ℉. Within the 200 ℉ to 150 ℉ band, the valve response is proportional to the temperature change. Outside of this band, the valve ceases to respond because it has reached its upper or lower limit. The proportional band range divided by the setpoint adjustment span is an expression of the percentage of full-scale operation. In this case, the proportional band is 33. 3 percent, because 50 divided by 150 equals 0. 333 or 33. 3. Most proportional controllers have an adjustable proportional band that varies from a few percent up to several hundred percent. For our water heating system to be operational，the steam valve must be opened to some percentage of its span. To start the system，assume that the setpoint is adjusted to 180 ℉. This generally causes the steam valve to be fully open. A full charge of steam is applied to the tank heating coil. As the temperature of the water begins to rise，the steam valve begins to close. The percentage of valve opening is proportional to the difference between the setpoint value and the actual temperature of the tank water. As the temperature of the water moves closer to the setpoint value，the percentage of valve opening is reduced. At 180 ℉，the valve will be 40 percent open. This represents the amount of steam needed to maintain the water temperature at 180 ℉. In a functioning system，this valve setting is somewhat unpredictable. Such things as ambient temperature，heat consumption of the water，insulation of the tank，temperature of the water source，and the temperature of the steam source are some of the conditions that must be taken into account. As a rule，proportional control works well in systems where the process changes are quite small and slow.

Selected from "Industrial Process Control Systems"，Dale R. Patrick，Stephen W. Fardo，The Fairmont Press，Inc.，2009.

Words and Expressions

1. **sophisticated [səˈfistikeitid]** *a*. **复杂的**
2. **intermediate [ˌintəˈmiːdiət]** *a*. **中间的，中级的**
3. **interior [inˈtiəriə(r)]** *a*. **内部的，本质的**
4. **furnace [ˈfəːnis]** *n*. **火炉，熔炉**
5. **adjustment [əˈdʒʌstmənt]** *n*. **调整，调节**
6. **oscillate [ˈɔsileit]** *vi*. **振荡，摆动**
7. **prolong [prəˈlɔːŋ]** *vt*. **延长，拖延**
8. **precise [priˈsais]** *a*. **精确的**
9. **ratio [ˈreiʃiəu]** *n*. **比率，比例**
10. **predetermined [ˌpridiˈtəːmind]** *a*. **先已决定的，预先确定的**
11. **coil [kɔil]** *n*. **线圈**
12. **insulation [ˌinsəˈleiʃən]** *n*. **绝缘，隔离**

Unit 17 ● Feedback and Its Effects

The motivation of using feedback, illustrated by the examples in Section 1-1, is somewhat oversimplified. In these examples, the use of feedback is shown to be for the purpose of reducing the error between the reference input and the system output. However, the significance of the effects of feedback in control systems is more complex than is demonstrated by these simple examples. The reduction of system error is merely one of the many important effects that feedback may have upon a system. We show in the following sections that feedback also has effects on such system performance characteristics as stability, bandwidth, overall gain, disturbance, and sensitivity.

To understand the effects of feedback on a control system, it is essential that we examine this phenomenon in a broad sense. When feedback is deliberately introduced for the purpose of control, its existence is easily identified. However, there are numerous situations wherein a physical system that we normally recognize as an inherently nonfeedback system turn out to have feedback when it is observed in a certain manner. In general, we can state that whenever a closed sequence of cause-and-effect relationships exists among the variables of a system, feedback is said to exist①. The viewpoint will inevitably admit feedback in a large number of systems that ordinarily would be identified as nonfeedback systems. However, with the availability of the feedback and control system theory, this general definition of feedback enables numerous systems, with or without physical feedback, to be studied in a systematic way once the existence of feedback in the sense mentioned previously is established.

We shall now investigate the effects of feedback on the various aspects of system performance. Without the necessary mathematical foundation of linear-system theory, at this point we can only rely on simple static-system notation for our discussion. Let us consider the simple feedback system configuration shown in Fig. 17. 1, where r is the input signal, y the output signal, e the error, and b the feedback signal. The parameters G and H may be considered as constant gains. By simple algebraic manipulations, it is simple to show that the input-output relation of the system is

$$M=\frac{y}{r}=\frac{G}{1+GH} \tag{17-1}$$

Using this basic relationship of the feedback system structure, we can uncover some of the significant effects of feedback.

Effect of Feedback on Overall Gain

As seen from Eq. (17-1), feedback affects the gain G of a nonfeedback system by a factor $1+GH$②. The system of Fig. 17. 1 is said to have negative feedback, since a minus sign is assigned to the feedback signal. The quantity GH may itself include a minus sign, *so the*

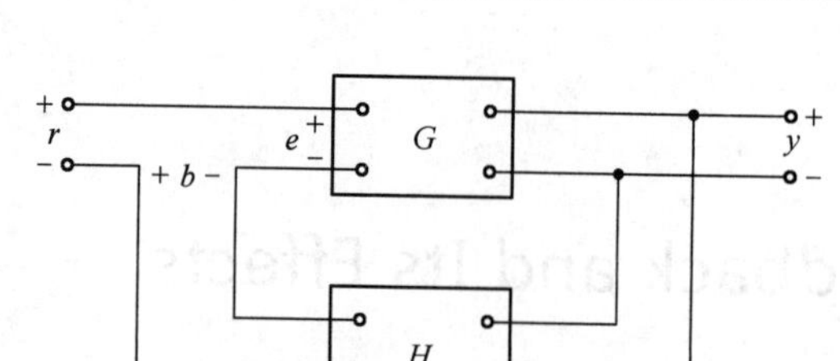

Fig. 17. 1 Feedback system

general effect of feedback is that it may increase or decrease the gain G. In a practical control system, G and H are functions of frequency, so the magnitude of $1+GH$ may be greater than 1 in one frequency range but less than 1 in another. Therefore, *feedback could increase the system gain in one frequency range but decrease it in another*.

Effect of Feedback on Stability

Stability is a notion that describes whether the system will be able to follow the input command, or be useful in general. In a nonrigorous manner, *a system is said to be unstable if its output is out of control*. To investigate the effect of feedback on stability, we can again refer to the expression in Eq. (17-1). If $GH=-1$, the output of the system is infinite for any finite input, and the system is said to be unstable. Therefore, we may state that *feedback can cause a system that is originally stable to become unstable*. Certainly, feedback is a two-edged sword; when used improperly, it can be harmful. It should be pointed out, however, that we are only dealing with the static case here, and in general, $GH=-1$ is not the only condition for instability.

It can be demonstrated that one of the advantages of incorporating feedback is that it can stabilize an unstable system. Let us assume that the feedback system in Fig. 17. 1 is unstable because $GH=-1$, if we introduce another feedback loop through a negative feedback gain of F, as shown in Fig. 17. 2, the input-output relation of the overall system is.

$$\frac{y}{r}=\frac{G}{1+GH+GF} \tag{17-2}$$

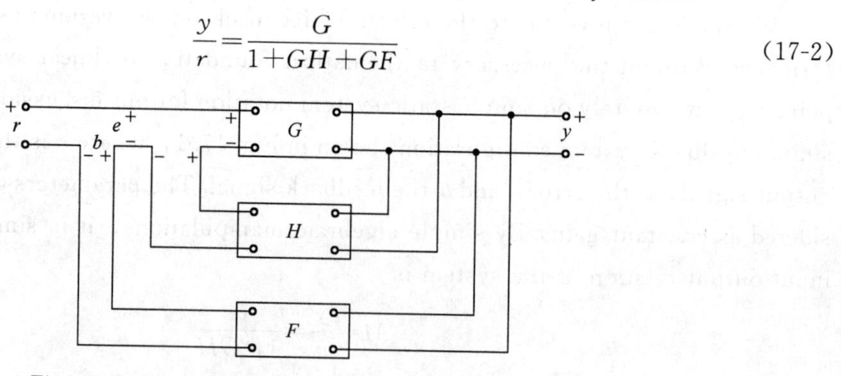

Fig. 17. 2 Feedback system with two feedback loops

It is apparent that although the properties of G and H are such that the inner-loop feedback system is unstable, because $GH=-1$, the overall system can be stable by proper selection of the outer-loop feedback gain F. In practice, GH is a function of frequency, and the stability condition of the closed-loop system depends on the magnitude and phase of GH. The bottom line is that *feedback can improve stability or be harmful to stability if it is not applied properly.*

Effect of Feedback on Sensitivity

Sensitivity considerations often are important in the design of control systems. Since all physical elements have properties that change with environment and age, we cannot always consider the parameters of a control system to be completely stationary over the entire operating life of the system. For instance, the winding resistance of an electric motor changes as the temperature of the motor rises during operation. The electronic typewriter described in Section 1-1 sometimes may not operate normally when first turned on due to the still-changing system parameters during warm-up. This phenomenon is sometimes called "morning sickness". Most duplicating machines have a warm-up period during which operation is blocked out when first turned on.

In general, a good control system should be very insensitive to parameter variations but sensitive to the input commands③. We shall investigate what effect feedback has on the sensitivity to parameter variations. Referring to the system shown in Fig. 17. 1, we consider G to be a gain parameter that may vary. The sensitivity of the gain of the overall system, M, to the variation in G is defined as

$$S_G^M=\frac{\partial M/M}{\partial G/G}=\frac{\text{percentage change in } M}{\text{percentage change in } G} \tag{17-3}$$

Where ∂M denotes the incremental change in M due to the incremental change in G, ∂G. By using Eq. (17-1), the sensitivity function is written

$$S_G^M=\frac{\partial M}{\partial G}\frac{G}{M}=\frac{1}{1+GH} \tag{17-4}$$

This relation shows that if GH is a positive constant, the magnitude of the sensitivity function can be made arbitrarily small by increasing GH, provided that the system remains stable. It is apparent that in an open-loop system, the gain of the system will respond in a one-to-one fashion to the variation in G (i. e. , $S_G^M=1$)。We again remind you that in practice, GH is a function of frequency; the magnitude of $1+GH$ may be less than unity over some frequency ranges, so that feedback could be harmful to the sensitivity to parameter variations in certain cases. In general, the sensitivity of the system gain of a feedback system to parameter variations depends on where the parameter is located. The reader can derive the sensitivity of the system in Fig. 17. 1 due to the variation of H.

Effect of Feedback on External Disturbance or Noise

All physical systems are subject to some types of extraneous signals or noise during operation. Examples of these signals are thermal-noise voltage in electronic circuits and brush or commutator noise in electric motors.

The effect of feedback on noise and disturbance depends greatly on where these extraneous signals occur in the system. No general conclusions can be reached, but in many situations, *feedback can reduce the effect of noise and disturbance on system performance.*

In general, feedback also has effects on such performance characteristics as bandwidth, impedance, transient response, and frequency response.

Selected from "Automatic Control Systems", Benjamin C. Kuo, Prentice-Hall, Inc., 1995

New Words and Expressions

1. bandwidth [ˌbændwidθ] *n.* 频带宽度，带宽，频宽
2. overall gain 总增益
3. impedance [imˈpiːdəns] *n.* 阻抗
4. turn out to 原来是
5. cause-and-effect relation 因果关系
6. inevitably [inˈevitəbli] *ad.* 不可避免地
7. notation [nəuˈteiʃən] *n.* （符号）表示法，标志法
8. algebraic [ˈældʒiˈbreiik] *a.* 代数学的，代数的
9. nonrigorous [ˈnɔnˈrigərəs] *a.* 不严格的，不严密的
10. extraneous [eksˈtreniəs] *a.* 外来的，非必要的，无关的
11. commutator [ˈkɔmjuteitə] *n.* 换向器，整流器
12. transient response 瞬态响应
13. two-edged sword 双刃剑，两面性

Notes

① 参考译文：通常可以这样说：只要系统变量间存在因果关系的封闭序列，就认为存在反馈。

② 参考译文：由方程式(17-1) 可知，反馈是通过因数 $1+GH$ 而影响非反馈系统的增益 G。

③ 参考译文：通常，一个良好的控制系统对于这些参数的变化应当很不敏感，而仍然能灵敏地跟踪指令。

Exercises

1. After reading the text above, write an abstract of it.
2. Answer the following questions according to the text.
   ① What did the author rely on for the discussion about the effects of feedback in this text? Why?
   ② Which aspects did the author show the effects of feedback on?
   ③ The feedback can be harmful, if it is improperly used, why?
   ④ What is a good control system?
3. Translate the 1st paragraph of the text into Chinese.
4. Put the following into Chinese by reference to the text.
   stability　bandwidth　overall gain　impedance　sensitivity　with a broad mind
   turn out to　cause-and-effect relation　without bound
5. Put the following into English.
   闭合序列　线性系统理论　静态系统　恒定增益　参考反馈　电动机绕组
6. Translate the following sentences into English.

① 通常，反馈系统的增益对于参数变化的灵敏度取决于参数所在的位置。

② *GH* 量本身也可能包含一个负号，所以反馈的总效应可以是增加增益，也可以是减小增益。

Reading Material 17

Types of Feedback Control Systems

Feedback control systems may be classified in a number of ways，depending on the purpose of the classification. For instance，according to the method of analysis and design，control systems are classified as linear and nonlinear，time-varying or time-invariant. According to the types of signal used in the system，reference is often made to continuous-data and discrete-data systems，or modulated and unmodulated systems. Control systems are often classified according to the main purpose of the system. For instance，a position-control system and a velocity-control system control the output variables according to the way the names imply. The type of a control system is defined according to the form of the open-loop transfer function. In general，there are many other ways of identifying control systems according to some special features of the system. It is important that some of these more common ways of classifying control systems are known so that proper perspective is gained before embarking on the analysis and design of these systems.

Linear Versus Nonlinear Control Systems

This classification is made according to the methods of analysis and design. Strictly speaking，linear systems do not exist in practice，since all physical systems are nonlinear to some extent. Linear feedback control systems are idealized models fabricated by the analyst purely for the simplicity of analysis and design. When the magnitudes of the signals in a control system are limited to a range in which system components exhibit linear characteristics (i. e.，the principle of superposition applies)，the system is essentially linear. But when the magnitudes of signals are extended beyond the range of the linear operation，depending on the severity of the nonlinearity，the system should no longer be considered linear. For instance，amplifiers used in control systems often exhibit a saturation effect when their input signals become large；the magnetic field of a motor usually has saturation properties. Other common nonlinear effects found in control systems are the backlash or dead play between coupled gear members，nonlinear spring characteristics，nonlinear friction force or torque between moving members，and so on. Quite often，nonlinear characteristics are intentionally introduced in a control system to improve its performance or provide more effective control. For instance，to achieve minimum-time control，an on-off (bang-bang or relay) type of controller is used in many missile or spacecraft control systems. Typically in these systems，jets are mounted on sides of the vehicle to provide reaction torque for attitude control. These jets are often controlled in a full-on or full-off fashion，so a fixed amount of air is applied from a given jet for a certain time period to control the attitude of the space vehicle.

For linear systems, there exists a wealth of analytical and graphical techniques for design and analysis purposes. A majority of the material in this book is devoted to the analysis and design of linear systems.

Time-Invariant Versus Time-Varying Systems

When the parameters of a control system are stationary with respect to time during the operation of the system, the system is called a time-invariant system. In practice, most physical systems contain elements that drift or vary with time. For example, the winding resistance of an electric motor will vary when the motor is first being excited and its temperature is rising. Another example of a time-varying system is a guided-missile control system in which the mass of the missile decreases as the fuel on board is being consumed during flight. Although a time-varying system without nonlinearity is still a linear system, the analysis and design of this class of systems are usually much more complex than that of the linear time-invariant systems.

Continuous-Data Control Systems

A continuous-data system is one in which the signals at various parts of the system are all functions of the continuous time variable t. Among all continuous-data control systems, the signals may be further classified as AC or DC. Unlike the general definitions of AC and DC signals used in electrical engineering, AC and DC control systems carry special significance in control systems terminology. When one refers to an ac control system, it usually means that the signals in the system are *modulated* by some form of modulation scheme. On the other hand, when a DC control system is referred to, it does not mean that all the signals in the system are *unidirectional*; then there would be no corrective control movement. A DC control system simply implies that the signals are *unmodulated*, but they are still AC signals according to the conventional definition. The schematic diagram of a closed-loop DC control system is shown in Fig. 17. 3. Typical waveforms of the signals in response to a step-function input are shown in the figure. Typical components of a DC control system are potentiometers, DC amplifiers, DC motors, and DC tachometers, and so on.

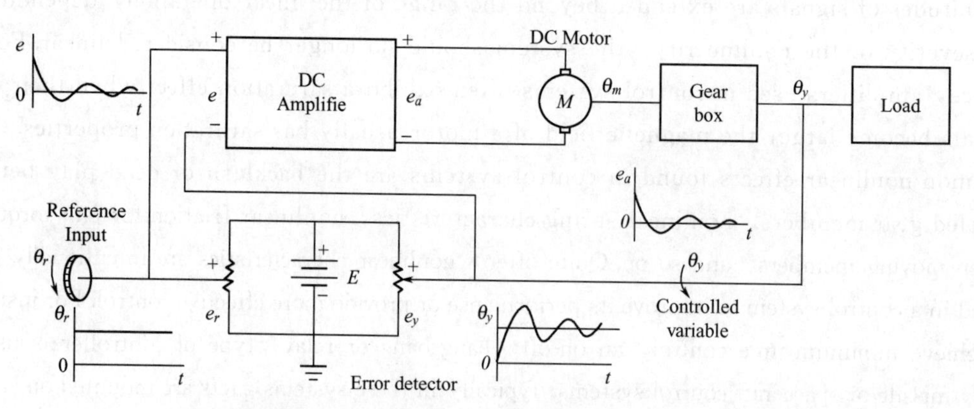

Fig. 17. 3 Schematic diagram of a typical DC closed-loop system

The schematic diagram of a typical AC control system that performs essentially the same task as that in Fig. 17. 4 is shown in Fig. 17. 4. In this case, the signals in the system are modulated; that is, the information is transmitted by an AC carrier signal. Notice that the output-controlled variable still behaves similarly to that of the DC system. In this case, the modulated signals are demodulated by the low-pass characteristics of the AC motor. AC control systems are used extensively in aircraft and missile control systems, in which noise and disturbance often create problems. By using modulated AC control systems with carrier frequencies of 400 Hz or higher, the system will be less susceptible to low-frequency noise. Typical components of an AC control system are synchros, AC amplifiers, AC motors, gyroscopes, accelerometers, and so on.

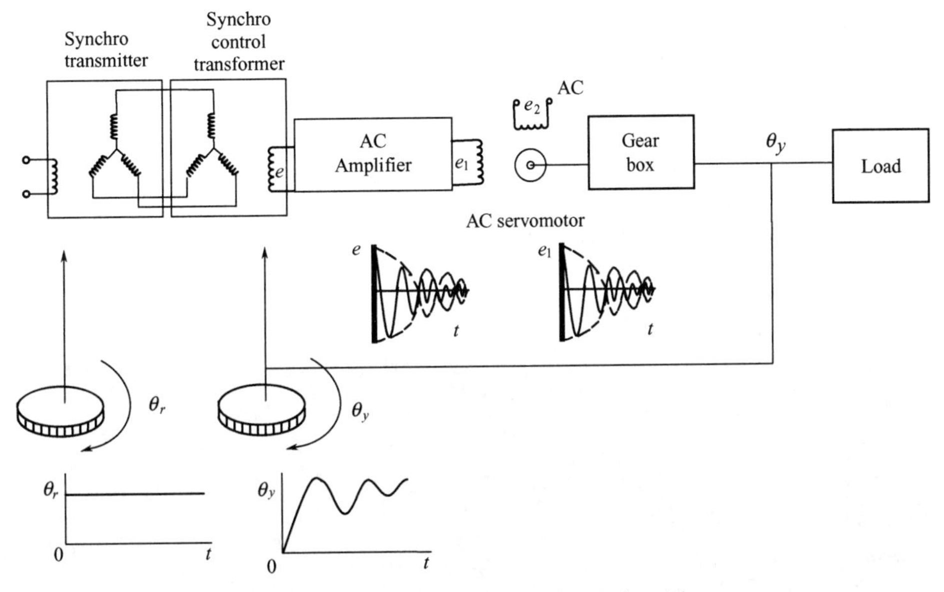

Fig. 17. 4 Schematic diagram of a typical AC closed-loop system

In practice, not all control systems are strictly of the AC or the DC type. A system may incorporate a mixture of AC and DC components, using modulators and demodulators to match the signals at various points of the system.

Discrete-Data Control Systems

Discrete-data control systems differ from the continuous-data systems in that the signals at one or more points of the system are in the form of either a pulse train or a digital code. Usually, discrete-data control systems are subdivided into sampled-data and digital control systems. Sampled-data control systems refer to a more general class of discrete-data control systems in which the signals are in the form of pulse data. A digital control system refers to the use of a digital computer or controller in the system, so that the signals are digitally coded, such as in binary code. For example, the printwheel control system is a typical digital control system, since the microprocessor receives and outputs digital data.

In general, a sampled-data system receives data or information only intermittently at specific instants of time. For instance, the error signal in a control system can be supplied only in the form of pulses, in which case the control system receives no information about the error signal during the periods between two consecutive pulses.

Selected from "Automatic Control Systems", Benjamin C. Kuo, Prentice-Hall, Inc., 1995.

New Words and Expressions

1. time-varying　时变的
2. time-invariant　定常的
3. continuous-data　连续数据
4. discrete [di'skriːt] *a*. 离散的，分立的，不连续的
5. unmodulated [ʌn'mɔdjuleitid] *a*. 未调制的，非调制的，未经调制的
6. saturation ['sætʃə'reiʃən] *n*. 饱和
7. potentiometer [pəˌtenʃi'ɔmətə] *n*. 电位计，电势计
8. synchro ['siŋkrəu] *n*. 自整角机，同步机
9. consecutive [kən'sekjutiv] *a*. 连续的，连贯的

Unit 18 • Adaptive Control

The most recent class of control techniques to be used are collectively referred to as adaptive control. Although the basic algorithms have been known for decades, they have not been applied in many applications because they are calculation-intensive. However, the advent of special-purpose digital signal processor (DSP) chips has brought renewed interest in adaptive-control techniques. The reason is that DSP chips contain hardware that can implement adaptive algorithms directly, thus speeding up calculations.

The main purpose of adaptive control is to handle situations where loads, inertias, and other forces acting on the system change drastically.

A classic example of a system with changing parameters is a guided missile. Missile mass drops as fuel burns, and it encounters differing friction at different altitudes.

Some system changes can be unpredictable, and ordinary closed-loop systems may not respond properly when the system transfer function varies. Sometimes, these effects can be handled by conventional linear-control techniques such as gain scheduling (feed-forward control). Conservative design practices may also enable some systems to remain stable even when subjected to parameter changes or unanticipated disturbances.

The price paid for such stability is suboptimal performance, however. Response to changes may be sluggish. Errors may fail to stay within satisfactory limits, or designs must compensate for loose error tolerances in other ways.

Adaptive control can help deliver both stability and good response. The approach changes the control algorithm coefficients in real time to compensate for variations in the environment or in the system itself. In general, the controller periodically monitors the system transfer function and then modifies the control algorithm. It does so by simultaneously learning about the process while controlling its behavior. The goal is to make the controller robust to a point where the performance of the complete system is as insensitive as possible to modeling errors and to changes in the environment[①].

Even ordinary feedback-control systems are adaptive in a limited sense, in that they can compensate for changes at their input that are within the system bandwidth. But these changes are comparatively small. Such systems can become unstable for large input swings, or may simply be unable to compensate for sufficiently large input changes.

There are two main approaches to adaptive feedback-control design: model reference adaptive control (MRAC) and self-tuning regulators (STRs). In MRAC, a reference model describes system performance. The adaptive controller is then designed to force the system or plant to behave like the reference model. Model output is compared to the actual output, and the difference is used to adjust feedback controller parameters.

Most work on MRAC has focused on the design of the adaptation mechanism. This mechanism must note the output error and determine how to adjust the controller coefficients. It must also remain stable under all conditions. One problem with the approach is that there is no general theoretical method of designing an adapter. Thus, most adapter functions are specially keyed to some kind of end application.

An advantage of MRAC is that it provides quick adaptations for defined inputs. A disadvantage is that it has trouble adapting to unknown processes or arbitrary disturbances.

Model-reference controllers have an adaptation mechanism. The comparable component in self-tuning regulators is a tuning algorithm. A self-tuning regulator assumes a linear model for the process being controlled (which is generally nonlinear). It uses a feedback-control law that contains adjustable coefficients. Self-tuning algorithms change the coefficients.

These controllers typically contain an inner and an outer loop. The inner loop consists of an ordinary feedback loop and the plant. This inner loop acts on the plant output in conventional ways. The outer loop adjusts the controller parameters in the inner feedback loop. The outer loop consists of a recursive parameter estimator combined with a control design algorithm.

The recursive estimator monitors plant output and estimates plant dynamics by providing parameter values in a model of the plant. These parameter estimates go to a control-law design algorithm that sends new coefficients to the conventional feedback controller in the inner loop.

The above description tends to be abstract because many different types of controllers and schemes are used to estimate parameters. Among the most widely used controllers are PID state controllers, and deadbeat controllers. Recursive parameter estimation techniques include stochastic approximation, least squares, extended Kalman filtering, and the maximum likelihood method.

Selected from Wikipedia, the free encyclopedia. http://www.reference.com

New Words and Expressions

1. digital signal processor (DSP) 数字信号处理器
2. inertia [i'nəːʃjə] *n.* 惯性，惯量，惰性，惰力，不活泼
3. transfer function 传递函数
4. gain scheduling 增益调度，增益调节，增益规划
5. feed-forward control 前馈控制
6. suboptimal ['sʌb'ɔptiməl] *a.* 次优的，未达最佳标准的，不最适宜的
7. sluggish ['slʌgiʃ] *a.* 惰性的；黏滞的；不活泼的；小［低］灵敏度的
8. input swing 输入摆幅
9. model reference adaptive control (MRAC) 模型参考自适应控制
10. self-tuning regulator (STR) 自校正调节器
11. deadbeat controller 无差拍控制器

12. recursive [ri'kəːsiv] *a*. 递归的，循环的
13. stochastic [stəu'kæstik] *a*. 随机的，机遇的，不确定的，概率性的
14. stochastic approximation 随机逼近（法）
15. least square 最小平方，最小二乘法
16. filter ['filtə] *n*.；*v*. 滤波器，过滤器；筛选程序；过滤，用过滤法除去
17. extended Kalman filtering 扩展卡尔曼滤波
18. maximum likelihood method 最大似然方法

Notes

① 参考译文：目的是令控制器牢靠，以便使整个系统性能对建模误差和环境变化尽可能不敏感。

Exercises

1. After reading the text above，write an abstract of it.
2. Answer the following questions according to the text.
   ① What is the main purpose of adaptive control? Please give an example.
   ② What is the characteristic of an adaptive control?
   ③ How does a model reference adaptive control work?
   ④ What is the difference between model reference adaptive control and self-tuning regulators?
3. Translate the 6th paragraph of the text into Chinese.
4. Put the following into Chinese by reference to the text.

   adaptive control　gain scheduling　real time　input swing
   plant　transfer function　estimator
5. Put the following into English.

   模型参考自适应控制　自校正调节器　前馈控制　惯性
   对……不灵敏　无差拍控制器　最小二乘法　随机的
6. Translate the following sentences into English.
   ① 当作用于系统的载荷、惯性力以及其他力发生急剧变化时，自适应控制能对其进行控制。
   ② 系统的有些变化是不可预测的，当系统的传递函数发生变化时，普通闭环系统可能无法作出适当的响应。

Reading Material 18

Digital Control Development

With the decreasing cost of digital hardware，economical digital control implementation is now feasible. Such applications include process control，automatic aircraft stabilization and control，guidance and control of aerospace vehicles，robotic control，and numeric control of manufacturing machines. An extensive example is state-feedback control systems for which the computer is used to estimate the inaccessible states and to help minimize parameter variations. The development of digital control systems can be illustrated by the following examples, which center on digital flight control system, process control systems, and robotics.

Aircraft Digital Control Development.

Modern technology has brought about some numerous changes in aircraft flight control systems. A step in the evolution of flight control systems is the use of a fly-by-wire (FBW) control system. In this design all pilot commands are transmitted to the control-surface actuators through electric wires. Individual component reliability was also increased by replacing the older analog circuitry with newer digital hardware. This updated system is referred to as a digital flight control system (DFCS). The use of airborne digital processors further reduced the cost, weight, and maintenance of modern aircraft and spacecraft. Other advantages associated with newer digital equipment included greater accuracy, increased modification flexibility through the use of software chances, improved dynamic reconfiguration techniques, and more reliable preflight and postflight maintenance testing.

Process Control

Analog computers were initially utilized to monitor and control small dimensional processes. Specific installations may have used regulation control (set-point), supervisory control (monitor), or on-line instruction. In directed digital control (DDC) the digital computer measures directly the set-point values and current outputs in order to generate control signals (value position, vehicle turning rate, etc.) for tracking or regulation. With supervisory control, the objective is to optimize a multiprocess system by measuring system parameters and generating the required controls. Examples include high-profit manufacturing facilities, very accurate autopilots, and efficient chemical processes. For large systems, a hierarchical supervisory structure provides for high-level scheduling, error recovery, and information management, and uses DDCs for controlling low-level individual plants.

Because of the increasing dimensionality of processes along with accuracy specifications, minicomputers and microcomputers are used to implement efficiently DDCs and supervisory control systems. The added flexibility and general-purpose display capabilities of a digital-computer system associated with micro-miniaturization [very large scale integrated circuit (VLSI) technology] have evolved contemporary process control systems with considerable distributed processing. These physically small multicomputer systems are used in modern manufacturing as integrated components of automobile processes, appliance sequencing, medical operations, and remote environmental monitoring and control systems.

Robotic Arm

A specific example of a process control function is the use of assembly robots. Fig. 18. 1 depicts a general-purpose robot arm that can be employed on a production line for assembling parts, mixing combustible chemicals, or inspecting components in an austere environment. Although the various arm joints are coupled through the structural links, an initial approach is to control each joint separately. Each arm joint can be rotated in a two-dimensional plane by a motor/gear system. Other arms may use direct drive motors. A coupled-joint

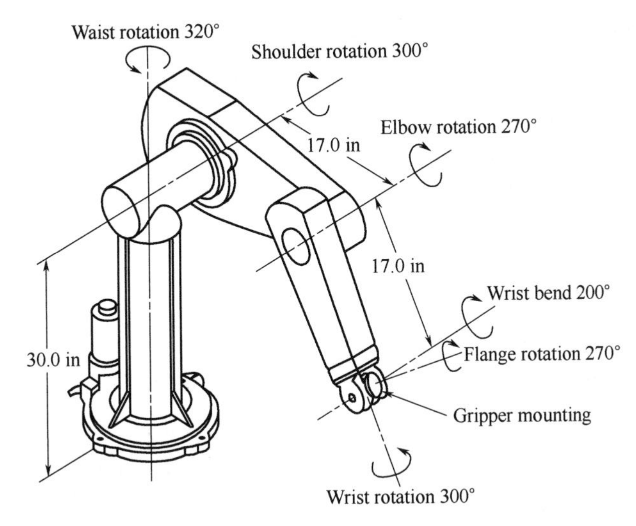

Fig. 18. 1 Example of robot arm (PUMA 600 Series)

robot-arm-controller design usually requires linearizing a set of nonlinear coupled equations in terms of position and velocity variables. At a supervisory level, the trajectories of the arm are defined in a software routine so that the desired movements can occur. Because of the power of large robots, the supervisory level also requires measurements of the environment so that destructive activity does not happen. Most position control systems historically have used a feedback combination of position, integral and derivative or rate signals (PID controller).

Nature of the Engineering Control Problem

In general, the engineering solution to a digital control problem can be decomposed into the following steps:

① Establish a set of performance specifications relating system input to output based upon given criteria (tracking response, disturbance rejection, sensitivity to plant variations/uncertainty, energy utilization, etc.).

② Generate a linear model (difference or differential equations) of the plant that describes the basic (or original) physical system. The model is usually a linear approximation to the real world. This effort is critical in understanding the process model.

③ As a result of system analysis or testing to desired performance specifications, a control problem exists and must be solved. Determine the performance of the basic system model by application of the available methods of control-theory analysis in the time domain or frequency domain.

④ Augment the system model with appropriate sensor and actuator dynamic models that respectively measure or control the desired plant parameters. Proper selection of measurement and actuator transducers is also of primary importance.

⑤ Use conventional control-theory, quantitative feedback theory (QFT), eigenstructure assignment, or modem control theory design techniques to design and synthesize

digital controllers to meet system performance specifications.

⑥ Simulate the overall control system, iterating the design until performance specifications are achieved.

⑦ Emulate, test, and iterate the implemented design. Technical knowledge of contemporary digital-computer hardware and software architectures has a critical influence on cost and real-time operation.

Design of the system to obtain the desired performance is the control problem. The necessary basic equipment is then assembled into a system to perform the desired control function. To a varying extent, most systems are nonlinear. In many cases the nonlinearity is small enough to be neglected, or the limits of operation are small enough to allow a linear analysis to be made. In this textbook linear systems or those which can be approximated as linear systems are considered. Because of the relative simplicity and straightforwardness of this approach, the reader can obtain a thorough understanding of linear systems. After mastering the terminology, definitions, and methods of analysis for linear control systems, the engineer will find it easier to undertake a study of nonlinear systems.

A basic system has the minimum amount of equipment necessary to accomplish the control function. The differential and difference equations that describe the physical system are derived and an analysis of the basic system is made. If the analysis indicates that the desired performance has not been achieved with this basic system, additional equipment must be inserted into the system or new control algorithms employed. Generally, this analysis also indicates the characteristics for the additional equipment or algorithms that are necessary to achieve the desired performance. After the system is synthesized to achieve the desired performance, based upon linear analysis, final adjustments can be made on the actual system to take into account the nonlinearities that were neglected. For digital control systems it is necessary to use good structured programming techniques and to document on a continuing basis all aspects of the software development.

Selected from "Digital Control Systems, Theory, Hardware, Software", Constantine H. Houpis, Gary B. Lamont, McGraw-Hill, Inc. 1992.

New Words and Expressions

1. implementation [ˌimplimenˈteiʃən] *n.*; *v.* 供给器具，装置，仪器；执行，实现
2. hierarchical [ˌhaiəˈrɑːkikəl] *a.* 分等级的，分级，分层，层次
3. assembly line 装配线
4. assembly parts 装配［组装］件，组合零件
5. combustible [kəmˈbʌstəbl] *a.*; *n.* 易［可］燃的；可燃物
6. austere [ɔsˈtiə] *a.* 严峻的；苛刻的
7. integral [ˈintigrəl] *a.* 整体的，完整的；总体的，总和的；必备的；积分的，累积的
8. derivative [diˈrivətiv] *a.*; *n.* 导出［生］的，派生的；［数］导数，微商

9. augment [ɔːɡment] *v.*；*n.* 增加［大、进］，扩张［大］；增加
10. eigenstructure　本征结构
11. synthesize [ˈsinθisaiz] *v.*（人工）合成，制造；综合（处理）
12. emulate [ˈemjuleit] *v.* 模仿，仿真
13. iterate [ˈitəreit] *v.* 迭代重复，反复法
14. terminology [ˌtəːmiˈnɔlədʒi] *n.* 专门名词，术语，词汇，术语学

Unit 19 • Artificial Intelligence (AI)

The essence of intelligence is learning. Just as humans learn how to communicate, identify visual patterns, or drive a car, machines can similarly be trained to perform such tasks based on powerful learning algorithms. Typical applications of AI include autonomous driving, computer vision, decision making, or natural language processing. AI holds the benefit of being adaptable to very heterogeneous contexts just like humans. Well-trained AI is capable of performing certain tasks at the same skill level as humans but with the additional advantages of high scalability and no need for pauses. AI can discover patterns in the data that are too complex for human experts to recognize. In some specific applications such as computer vision, AI has already achieved performance levels surpassing that of humans (e.g., in skin cancer diagnostics). The idea of AI dates back to the 1950s when AI successes were largely limited to the scientific field. Nowadays, adoption of AI has become increasingly easier due to freely available algorithms and libraries, relatively inexpensive cloud-based computing power, and the proliferation of sensors generating data. ①

In the industrial sector, AI application is supported by the increasing adoption of devices and sensors connected through the Internet of Things (IoT). Production machines, vehicles, or devices carried by human workers generate enormous amounts of data. AI enables the use of such data for highly value-adding tasks such as predictive maintenance or performance optimization at unprecedented levels of accuracy. Hence, the combination of IoT and AI is expected to kick off the next wave of performance improvements, especially in the industrial sector.

The nomenclature of artificial intelligence

Artificial intelligence is a buzzword these days and, hence, subject to multiple interpretations. For the purpose of establishing a common understanding, we have defined various AI terms as they are used in this report.

- Artificial intelligence (AI) is intelligence exhibited by machines, with machines mimicking functions typically associated with human cognition. AI functions include all aspects of perception, learning, knowledge representation, reasoning, planning, and decision making. The ability of these functions to adapt to new contexts, i.e., situations that an AI system was not previously trained to deal with, is one aspect that differentiates strong AI from weak AI.
- Machine learning (ML) describes automated learning of implicit properties or underlying rules of data. It is a major component for implementing AI since its output is used as the basis for independent recommendations, decisions, and feedback mechanisms. Machine learning is an approach to creating AI. As most AI systems today are based on ML, both terms are often used interchangeably-particularly in the business context.

- Machine learning uses training, i. e. , a learning and refinement process, to modify a model of the world. The objective of training is to optimize an algorithm's performance on a specific task so that the machine gains a new capability. Typically, large amounts of data are involved. The process of making use of this new capability is called inference. The trained machine-learning algorithm predicts properties of previously unseen data.
- There are three main types of learning within ML, namely supervised learning, reinforcement learning, and unsupervised learning . They differ in how feedback is provided. Supervised learning uses labeled data ("correct answer is given") while unsupervised learning uses unlabeled data ("no answer is given"). In reinforcement learning, feedback includes how good the output was but not what the best output would have been. In practice, this often means that an agent continuously attempts to maximize a reward based on its interaction with its environment.
- Since the late 2000s, deep learning has been the most successful approach to many areas where machine learning is applied. It can be applied to all three types of learning mentioned above. Neural networks with many layers of nodes and large amounts of data are the basis of deep learning. Each added layer represents knowledge or concepts at a level of abstraction that is higher than that of the previous one. Deep learning works well for many pattern recognition tasks without alterations of the algorithms as long as enough training data is available. Thanks to this, its uses are remarkably broad and range from visual object recognition to the complex board game "Go".

AI-enhanced predictive maintenance (AI application)

Context

Predictive maintenance aims at improving asset productivity by using data to anticipate machine breakdowns. A well-established and relatively simple method of recognizing failures early on is condition monitoring. The complexity of forecasting failure is often due to the enormous amount of possible influencing factors. Data sources can be manifold and depend on the scenario. E. g. , in engines, gear boxes, or air conditioning, analysis of sound can detect an anomaly in device operation. In switches, machines, and robots, vibrations can be measured and used to detect errors. Since new sensors and IoT devices can be integrated in production processes and operations, the availability of data increases drastically. AI-based algorithms are capable of recognizing errors and differentiating the noise from the important information to predict breakdowns and guide future decisions. ②

Approach

Machine-learning techniques examine the relationship between a data record and the labeled output (e. g. , failures) and then create a data-driven model to predict those outcomes. This technique helps recognize patterns from historical events and either predict future failures or prevent them based on learnings from specific breakdown root causes. Companies like Neuron Soundware use artificial auditory cortexes to simulate human sound interpretation, thus automating and improving the detection and identification of potential breakdown causes. ③ KONUX, one of last year's winners of McKinsey's "The Spark" award for digital in-

novation, uses sensors to detect anomalies. Its cloud-based AI system continuously learns from alerts to further improve the overall performance of the system and give recommendations for optimized maintenance planning and extended asset life cycles. Recent applications of machine learning also combine supervised learning with unsupervised learning and feature learning. This enables an automated classification of machine failure modes and also the identification of relevant features in the data, thereby enhancing expert domain knowledge. Both approaches greatly simplify the deployment of predictive maintenance systems while improving prediction accuracy. In addition to algorithmic advances, the use of a great variety of data sources beyond sensor outputs-such as maintenance logs, quality measurement of machine outputs, and, if applicable, external data sources such as weather data-enables prediction of events that were not possible to model before. Implementing AI-supported predictive maintenance takes, on average, six to eight weeks for pilot cases and several months for a full rollout. It may take longer if sensor development is involved.

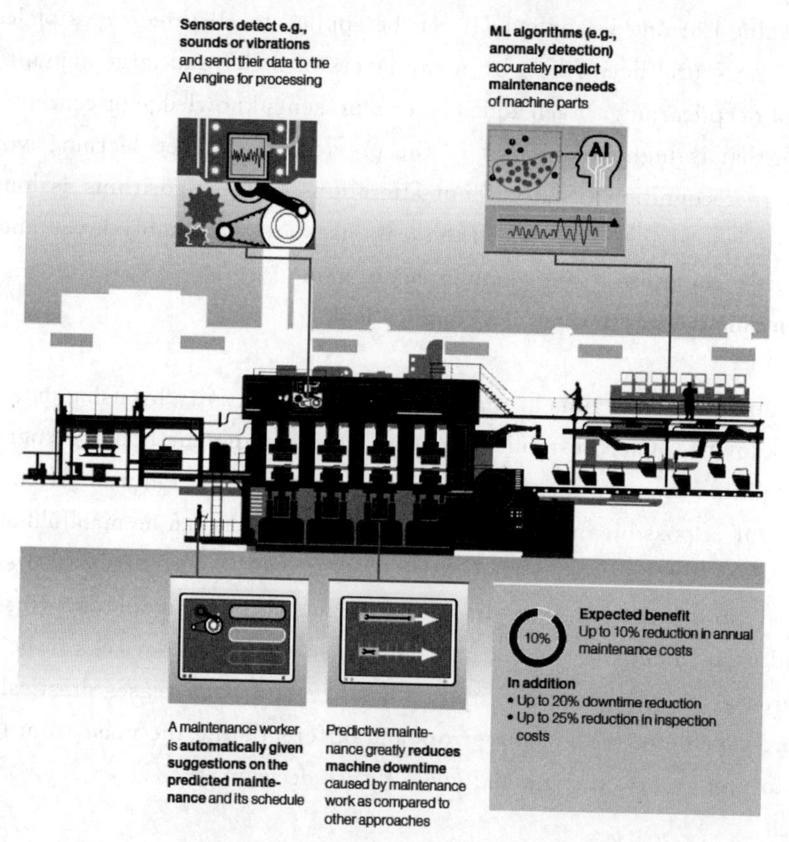

Fig. 19. 1 Schematic diagram of predictive maintenance

Impact

Comparing an AI-based approach to traditional condition monitoring or more classical maintenance strategies like usage-based exchange, a considerable improvement can be expected due to better failure prediction. Depending on the starting point and the level of redundancy, availability can sometimes increase by more than 20%. Inspection costs may be reduced by up to 25% and an overall reduction of up to 10% of annual maintenance costs is possible.

Selected from "Smartening up with Artificial Intelligence (AI)—What's in it for Germany and its Industrial Sector?" Copyright © McKinsey & Company, April 2017, www.mckinsey.com.

New Words and Expressions

1. algorithm [ˈælgəriðəm] *n*. 算法，计算程序
2. heterogeneous [ˌhetərəˈdʒiːniəs] *a*. 异类的，各种各样的
3. pause [pɔːz] *v*.; *n*. 暂停，停顿
4. proliferation [prəˌlifəˈreiʃən] *n*. 激增，涌现
5. internet of things (IoT) 物联网
6. nomenclature [nəˈmenklətʃər] *n*. 命名法，专门术语
7. mimicking [ˈmimikiŋ] *v*. 模仿（人的言行举止），（外表或行为举止）像，似
8. cognition [kɔgˈniʃən] *n*. 认知，感知，认识
9. perception [pərˈsepʃən] *n*. 知觉，感知，洞察力，看法，见解
10. implicit property 隐式属性
11. underlying rules 潜规则，潜在的规律
12. abstraction [æbˈstrækʃən] *n*. 抽象概念，抽象；提取；抽取；分离
13. manifold [ˈmænifəuld] *a*.; *vt*. 多的，多种多样的，许多种类的；增多，使…多样化
14. scenario [səˈnæriəu] *n*. 设想，方案，预测
15. anomaly [əˈnɔməli] *n*. 异常事物，反常现象
16. auditory cortex 听觉皮层
17. deployment [diˈplɔimənt] *n*. （部队、资源或装备的）部署，调集

Notes

① 参考译文：当今，由于免费提供的算法和数据库、相对廉价的云计算能力和生成数据的传感器的激增，采用人工智能变得越来越方便。

② 参考译文：基于人工智能的算法能够识别误差并从重要的信息中区分噪声来预测故障并指导未来的决策。

③ 参考译文：像 Neuron Soundware 公司使用人工听觉皮层来模拟人声解译，从而实现自动化并提高潜在故障原因的检测与识别能力。

Exercises

1. After reading the text above, write a summary of it.
2. Answer the following questions according to the text.
   ① What is the essence of intelligence?
   ② What are the additional advantages of the well-trained AI compared to humans?
   ③ What is the objective of training in machine learning?
   ④ How many main types of learning within machine learning? How are the learning types different from each other?
3. Translate the 1st paragraph into Chinese.
4. Put the following into Chinese by reference to the text.

implicit property　underlying rule　strong AI　supervised learning
reinforcement learning　unsupervised learning　auditory cortex

5. Put the following into English.
人工智能　自动驾驶　神经网络　预测性维修　维修日志
6. Put the following sentences into English.
① 深度学习对许多模式识别任务都很有效，只要有足够的训练数据，算法就不会改变。
② 预测性维修的目的是通过应用数据来预测机器故障从而提高生产能力。

Reading Material 19

Programmable Logic Controllers

Introduction

The modern programmable logic controller (PLC) is the successor of relay-based controls. The technological shift began in the 1960s, when the limitations of electromechanical relay-based controllers drove General Motors to search for electronic alternatives. The answer was provided in 1970 by Modicon, who provided a microprocessor-based control system. Its programming language was modeled after relay ladder logic diagrams to ease the transition of designers, builders, and maintainers to these new controllers. Throughout the 1970s the technology was refined and proven, and since the early 1990s they have become ubiquitous on the factory floor. These early controllers, or Programmable Logic Controllers (PLC), represented the fast systems that (1) could be used on the factory floor, (2) could have there "logic" changed without extensive rewiring or component changes, and (3) were easy to diagnose and repair when problems occurred.

Control system

All programmable controllers consist of the basic functional blocks shown in Fig. 19.2. We will examine each block to understand the relationship to the control system. First we look at the center, as it is the heart of the system. It consists of a microprocessor, logic memory for the storage of the actual control logic, storage or variable memory for use with data that will ordinarily change as a function of the control program execution, and a power supply to provide electrical power for the processor and memory. Next comes the I/O block. This function takes the control level signals for the CPU and converts them to voltage and current levels suitable for connection with factory grade sensors and actuators. The I/O type can range from digital, analog, or a variety of special purpose "smart" I/O which are dedicated to a certain application task. The programmer is normally used only to initially configure and program a system and is not required for the system to operate. It is also used in troubleshooting a system, and can prove to be a valuable tool in pinpointing the exact cause of a problem. The field devices shown here represent the various sensors and actuators connected to the I/O. These are the arms, legs, eyes, and ears of the system, including pushbuttons, limit switches, proximity switches, photo sensors, thermocouples, position sensing devices, and bar code reader as input; and pilot light, display devices, motor starters, DC and AC drivers, solenoids, and printers as outputs.

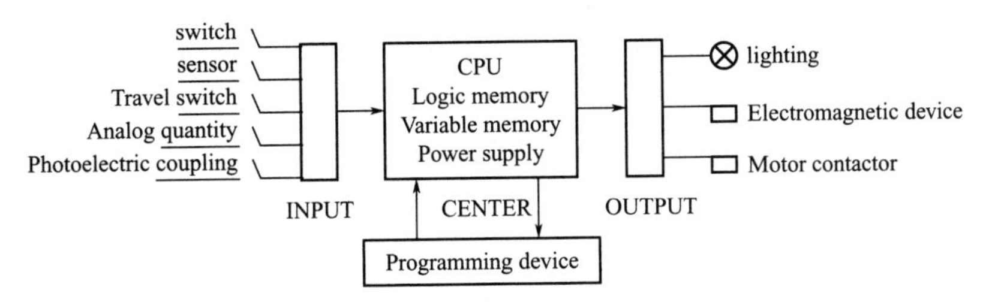

Fig. 19. 2　PLC control system

Programming Languages

Every PLC can be programmed with ladder logic. Ladder logic uses input contacts (shown with two vertical lines) and output coils (shown with a circle). A contact with a slash through it represents a normally closed contact. In ladder logic, the left-hand rail is energized. When the contacts are closed in the right combinations, power can flow through the coil to the right-hand neutral rail.

Consider the ladder logic example in Fig. 19. 3. It is assumed that the hot rail at the left side has power, and the right side rail is neutral. When the contacts are opened and closed in the right combinations they allow power to flow through the output coils, thus actuating them. The program logic is interpreted by working from the left side of the ladder. In the first rung if A and D are on, the output X will be turned on. This can also be accomplished by turning B on, turning C off, and turning D on. In the second, the output Y will be on if X is on and A is on, or D is off. Notice that the branches behave as OR functions and the contacts in line act as an AND function. It is possible to write ladder logic rungs as Boolean equations, as shown on the right-hand side of the figure.

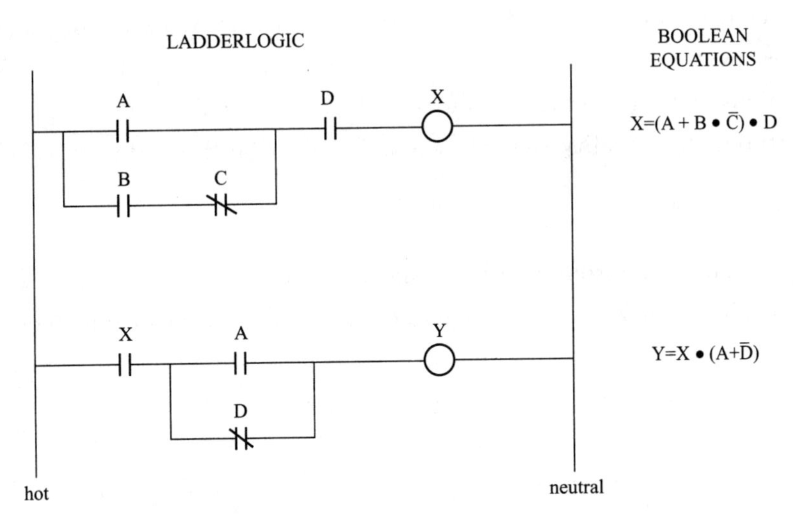

Fig. 19. 3　A simple ladder logic program with equivalent Boolean equations

The example in Fig. 19. 3 contains only conditional logic, but Fig. 19. 4 shows a more complex example of a ladder logic program that uses timers and memory values. When the **run** input is active, output heater will turn on, 5 s later fan 1 will turn on, followed by fan 2 at 10 s. The first rung of the program will allow the system to be started with a normally open run push button input, or stopped with a normally closed push button stop. All stop inputs are normally closed switches, so the contact in this rung needs to be normally open to reverse the logic. The output active is also used to branch around the run to seal-in the run state. The next line of ladder logic turns on an output **heater** when the system is active. The third line will run a timer when **active** is on. When the input to the TON timer goes on, the timer T4:0 will begin counting, and the timer element T4:0. ACC will begin to increment until the delay value of 10 s is reached, at this point the timer done bit T4:0/DN bit will turn on and stay on until the input to the timer is turned off. The fourth rung will compare the accumulated time of the timer and if it is greater than 5 the output fan 1 will be turned on. The final rung of the program will turn on fan 2 after the timer has delayed 10 s.

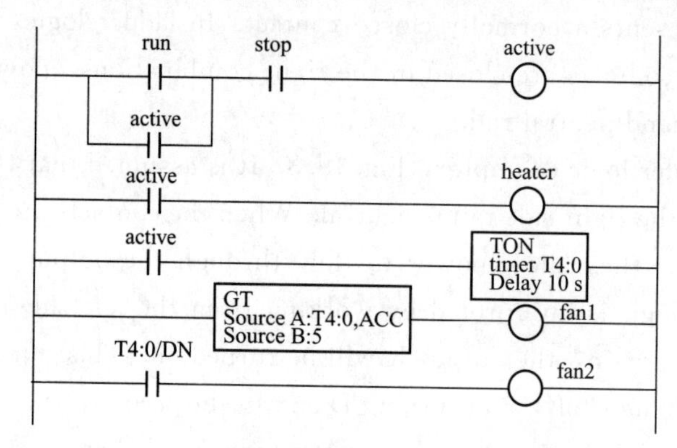

Fig. 19. 4　A complex ladder logic example

A PLC scans (executes) a ladder logic program many times per second. Typical execution times range from 5 to 100 ms. Faster execution times are required for processes operating at a higher speed.

The notations and function formats used in Fig. 19. 4 are based on those developed by a PLC manufacturer. In actuality, every vendor has developed a different version of ladder logic.

PLC Features

(a) High reliability, strong anti-jamming capability

High reliability is the key to performance of the electrical control equipment. PLC is due to the introduction of modern LSI technology, strict production process, the internal circuit to take the advanced anti-jamming technology, with high reliability. For example, the Mitsubishi F series PLC MTBF up to 300,000 hours. Redundant CPU PLC, the average time between failures is longer. From a machine outside the circuit of the PLC, the PLC constitutes a control system, and compared to the same scale relay contactor system, electrical wiring and switch contacts have been reduced to hundreds or even thousands, the fault will be greatly reduced. In addition, PLC with a self-detection of hardware failure, failure to timely alert. In the application software, application can also be incorporated into the peripheral devices and fault diagnostic procedures, also received the fault self-diagnostic protection circuit outside the PLC system and equipment. In this way, the entire system with high reliability is also not surprising.

(b) Complete, fully functional, the applicability

PLC development today, has formed a large, medium and small scale series products, and they can be used for industrial control applications of all sizes. In addition to the logical processing functions, the modern PLC mostly data computing power can be used for a variety of digital control. In recent years the emergence of a large number of functional units of the PLC, PLC penetrated into the position control, temperature control, CNC and other industrial control. With PLC communication capability enhancement and development of human-computer interface technology, the use of PLC composed of a variety of control systems have become very easy.

(c) Easy to use, well received by engineers and technicians

As a general industrial control computer, PLC is suitable for industrial control equipment of industrial and mining enterprises. Its interface is simple and programming language is easy to be accepted by engineers and technicians. The graphic symbols of ladder diagram language are quite close to the expression mode and relay circuit diagram. The function of relay circuit can be easily realized with only a few switching logic control instructions of PLC. It is convenient for people who are not familiar with electronic circuits, computer principles and assembly languages to use computers for industrial control.

(d) System design, construction workload is small, easy to maintain, easy to transform

PLC replaces the wiring logic with storage logic, which greatly reduces the external wiring of the control device and greatly shortens the design and construction cycle of the control system. More importantly, it is possible to change the production process of the same equipment through changing the procedures. This is very suitable for many varieties, small batch production occasions.

(e) Small size, light weight, low energy consumption

Ultra-Small PLC, for example, the newly produced species at the bottom of size less than 100mm, weighs less than 150g, only a few watts of power. Due to its small size, it is easy to fit inside the machine and is an ideal control device for mechatronics.

Selected from "Basic PLC programs", John Ridley, Elsevier Inc. 2003.

New Words and Expressions

1. programmable logic controller (PLC) 可编程控制器
2. alternative [ɔːl'təːnətiv] *a.*; *n.* 可供替代的；可供选择的事物
3. ladder logic diagrams 梯形逻辑图
4. extensive [ik'stensiv] *a.* 广泛的，大量的
5. limit switch 限位开关，行程开关

6. proximity [prɔk'siməti] *n*. 接近，邻近，靠近
7. proximity switch　接近开关
8. bar code reader　条码阅读器
9. thermocouple [θərməu'kʌpl] *n*. 热电偶
10. solenoid [sɔlənɔid] *n*. 螺线管
11. energize [enərdʒaiz] *v*. 为……提供电力（或能量），使通电
12. anti-jamming ['ænti 'dʒæmiŋ] *n*. 抗干扰
13. emergence [i'məːrdʒəns] *n*. 出现，兴起
14. penetrate ['penətreit] *v*. 穿过，进入，渗透

Unit 20 • Measurement Systems

The role of measurement is to provide information on system status which, in a mechatronic system, is used to control the operation of the system. This function is performed by the measurement modules which incorporate the necessary sensors and transducers together with any local signal processing[①].

Measurement technology is not a new science but can be traced back to origins in trade, where the need to quantify, on a basis which was both repeatable and representative, the nature of goods on offer led to the introduction of standards for measures such as volume and weight. With the growth in science and technology as exemplified by the Renaissance, more and specific measuring instruments evolved to support the growing interest in and investigation of the physical world.

With the advent of the industrial revolution, instrumentation and measurement science began to be applied to manufacturing. The early instruments were analogue in nature and provided information about the basic physical parameters concerned with the operation of the process. Feedback was achieved by the operator who read the instruments and then made the necessary adjustments to maintain the process within the required bounds[②].

The next step forward came with the introduction of the steam engine and automatic feedback control mechanisms such as the Watt governor for speed control. Following this lead, instrumentation gradually became more integrated with the process, providing feedback information to a controller which then regulated the system behavior with a minimum of manual intervention.

By the late 1960s the advantages of discrete control had been recognized and production processes were often controlled by relay based systems. Such relay controllers, though complex in configuration and layout, were relatively unsophisticated and incapable of handling complex information. The performance requirements of their associated sensors and transducers were correspondingly restricted to the supply of simple data at a level capable of being utilized by the controllers.

These conditions were largely maintained with the introduction of computer monitored, as opposed to controlled, systems in which the computer was used primarily in a supervisory role as a central machine receiving information sequentially from a number of radially connected outstations. With the advent of minicomputers such as the DEC PDP series, computer controlled systems increased in both number and scope, though they were limited in application by both processing power and cost. More recently, sophisticated microprocessor based programmable logic controllers together with distributed and embedded microprocessor systems have resulted in the complex interconnected designs that are now found in both products and manufacturing processes.

The growth in microelectronics and computing technologies from the mid 1970s that made available low cost processing power has resulted in a parallel growth in the demand for information about system conditions. In many cases this increased demand could not be met by existing sensors and transducers and resulted in the development of a range of advanced sensors and transducers, often incorporating local processing power, either integral with the sensing element on the same silicon slice or in the form of a single chip microprocessor.

Sensors, Transducers and Measurement

In any measurement system, sensors and transducers are used to provide information about system conditions. Unfortunately, the usage of the terms sensor and transducer is complicated by the variety of different meanings adopted internationally, including in some instances an apparent interchangeability. For the purpose of this book the following definitions are adopted:

Sensor

That part of the measurement system that responds to the particular physical parameter to be measured.

Transducer

That component of the system that transfers information in the form of energy from one part of the system to another, including in some cases changing the form of energy containing the information.

In addition to a sensor/transducer, a measurement may include a pre-processing stage and a post-processing stage, as shown in Fig. 20. 1. The pre-processing stage serves to characterize the information in the incoming signal before presentation to the sensing element. This then detects and responds to the physical stimulus at its input and provides the input to the transducer. The output signal from the transducer is then operated on by the post-processing stage to produce the final output. The relationship between these stages is illustrated by the optical encoder used for speed measurement shown in Fig. 20. 2. Here, the light from the source is pre-processed by the slotted disc to produce a series of pulses. These light pulses are received by the transducer which converts the signal into a series of electrical pulses. The post-processing stage then counts the number of pulses occurring in a defined time interval to determine the speed of rotation of the shaft carrying the encoder.

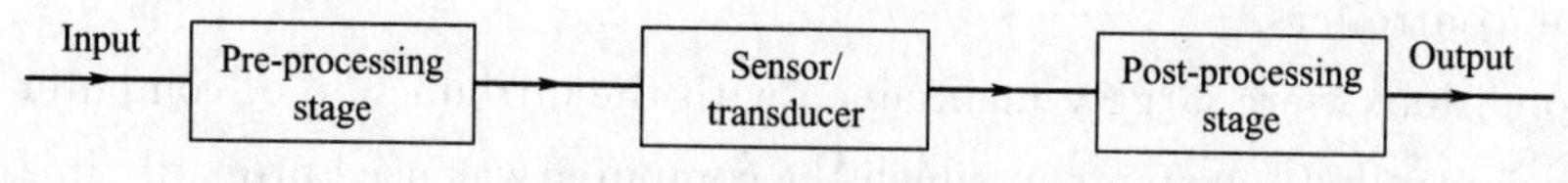

Fig. 20. 1 Pre- and post - processing of measurement data

As has been stated, the physical basis for the transmission of information by a transducer is that of energy transfer. This concept has been developed by Middelhoek to identify six signal energy domains for the transfer of information. These domains are as follows: radiant, mechanical, thermal, electrical, magnetic and chemical.

Of these, the electrical output domain is the most significant in a mechatronics context because of its information handling capability, and transducer outputs in other domains may

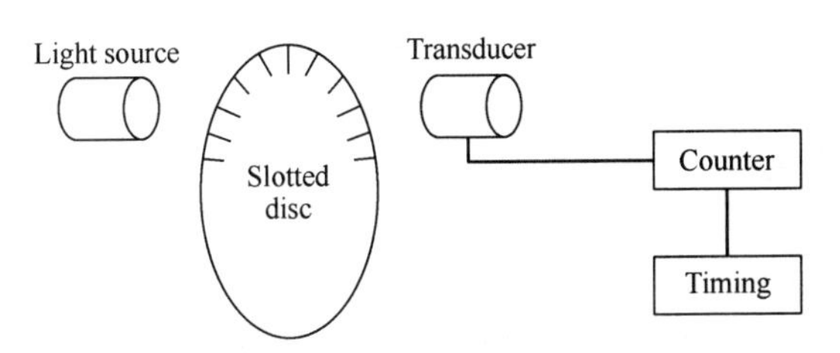

Fig. 20. 2　Speed measurement using a slotted disc with optical sensors and transducers

well be further processed to transfer the information into the electrical domain.

Selected from "Mechatronics: Electronics in Products and Processes", D. A. Bradley, D. Dawson, N. C. Burd and A. J. Loader, Chapman and Hall, 1991.

New Words and Expressions

1. module [ˈmɔdjuːl] *n*. 模块，组件，模件
2. trace back (…) to… (把……) 追溯到……
3. transducer [trænzˈdjuːsə] *n*. 变换器，换能器，传感器
4. renaissance [riˈneisəns] *n*. 文艺复兴
5. instrumentation [instrumenˈteiʃən] *n*. 仪表化
6. supervisory [sjuːpəˈvaizəri] *a*. 监督的，管理的
7. embed [imˈbed] *v*. 嵌入
8. interchangeability [intə(ː)ˈtʃeindʒəbiliti] *n*. 互换性，可交换性
9. pre-processing 前处理，前置处理
10. post-processing 后处理，后置处理
11. optical [ˈɔptikəl] *a*. 光学的，光导的
12. encoder [inˈkəudə] *n*. 编码器
13. domain [dəˈmein] *n*. 领域，范围，域，区域

Notes

① 参考译文：这一功能是由必要的传感器与转换器以及任何局部信号处理结合成的测量模块完成的。

incorporate M with N　　(把 M) 与 N 混合 [合并]

② 参考译文：反馈是通过操作人员读取仪器上的读数，然后做必要的调整以把工艺过程维持在所要求的范围之内来达到的。

Exercises

1. After reading the text above, summarize its main idea in oral English.
2. Answer the following questions according to the text.

① What is the role of measurement in a mechatronic system?

② What is the definition of sensor and transducer?

③ What are the components of a measurement system?

④ What is the physical basis for the transmission of information by a transducer? How many signal energy domains are there for the transfer of information? What are they?

3. Translate the 7th paragraph of the text into Chinese.
4. Put the following into Chinese by reference to the text.
   relay based system　　sensor　　transducer　　optical encoder　　domain
   be restricted to
5. Put the following into English.
   机电一体化系统　　前置处理　　后置处理　　互换性
   可重复的　　有代表性的　　单片机
6. Translate the following sentences into English.
   ① 随着工业革命的到来，仪表和测量科学被逐步用于制造业。
   ② 微电子技术和计算技术的进步使得对系统条件信息的需求相应增加。

Reading Material 20

Industrial Instrumentation and Process Control

Introduction

Instrumentation is the basis for process control in industry. However, it comes in many forms from domestic water heaters and HVAC, where the variable temperature is measured and used to control gas, oil, or electricity flow to the water heater, or heating system, or electricity to the compressor for refrigeration, to complex industrial process control applications such as used in the petroleum or chemical industry.

In industrial control a wide number of variables, from temperature, flow, and pressure to time and distance, can be sensed simultaneously. All of these can be interdependent variables in a single process requiring complex microprocessor systems for total control. Due to the rapid advances in technology, instruments in use today may be obsolete tomorrow, as new and more efficient measurement techniques are constantly being introduced. These changes are being driven by the need for higher accuracy, quality, precision, and performance. To measure parameters accurately, techniques have been developed that were thought impossible only a few years ago.

Process Control

In order to produce a product with consistently high quality, tight process control is necessary. A simple-to-understand example of process control would be the supply of water to a number of cleaning stations, where the water temperature needs to be kept constant in spite of the demand. A simple control block is shown in Fig. 20. 3(a), steam and cold water are fed into a heat exchanger, where heat from the steam is used to bring the cold water to the required working temperature. A thermometer is used to measure the temperature of the water (the measured variable) from the process or exchanger. The temperature is observed by an operator who adjusts the flow of steam (the manipulated variable) into the heat exchanger to keep the water flowing from the heat exchanger at the constant set temperature. This operation is referred to as process control, and in practice would be automated as shown in Fig. 20. 3(b).

Process control is the automatic control of an output variable by sensing the amplitude of the output parameter from the process and comparing it to the desired or set level and feeding an error signal back to control an input variable — in this case steam. See Fig. 20. 3 (b). A temperature sensor attached to the outlet pipe senses the temperature of the water flowing. As the demand for hot water increases or decreases, a change in the water temperature is sensed and converted to an electrical signal, amplified, and sent to a controller that evaluates the signal and sends a correction signal to an actuator. The actuator adjusts the flow of steam to the heat exchanger to keep the temperature of the water at its predetermined value.

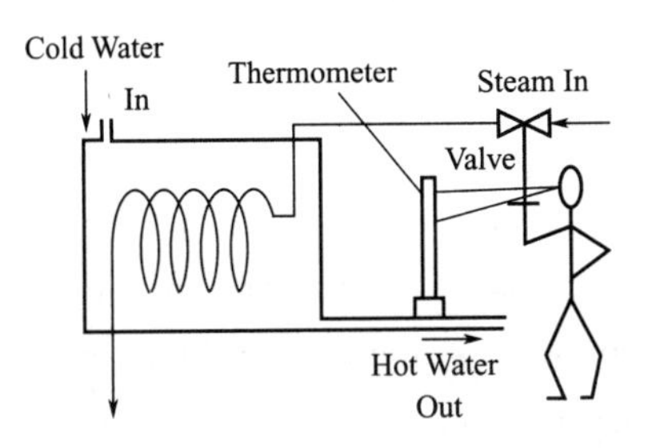

(a) Shows the manual control of a simple heat exchanger process loop

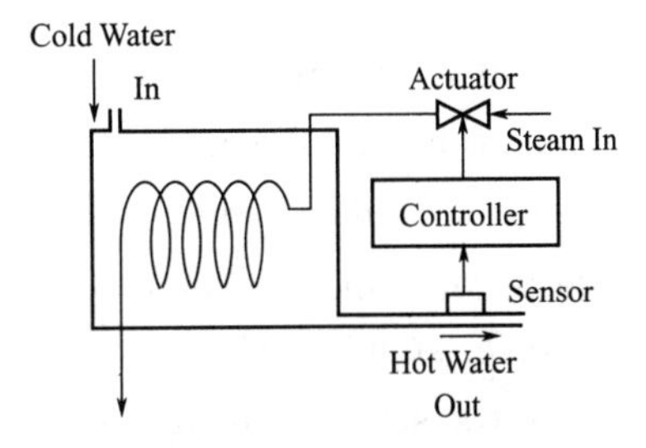

(b) Automatic control of a heat exchanger process loop

Fig. 20. 3 Process control

The diagram in Fig. 20. 3 (b) is an oversimplified feedback loop and is expanded in Fig. 20. 4. In any process there are a number of inputs, i. e. , from chemicals to solid goods. These are manipulated in the process and a new chemical or component emerges at the output. The controlled inputs to the process and the measured output parameters from the process are called variables.

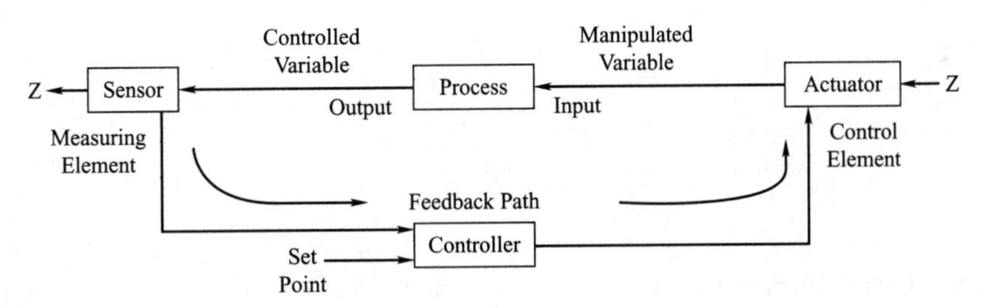

Fig. 20. 4 Block diagram of a process control loop

In a process-control facility the controller is not necessarily limited to one variable, but can measure and control many variables. A good example of the measurement and control of multivariables that we encounter on a daily basis is given by the processor in the automobile engine. Fig. 20. 5 lists some of the functions performed by the engine processor. Most of the controlled variables are six or eight devices depending on the number of cylinders in the engine. The engine processor has to perform all these functions in approximately 5 ms. This example of engine control can be related to the operations carried out in a process-control operation.

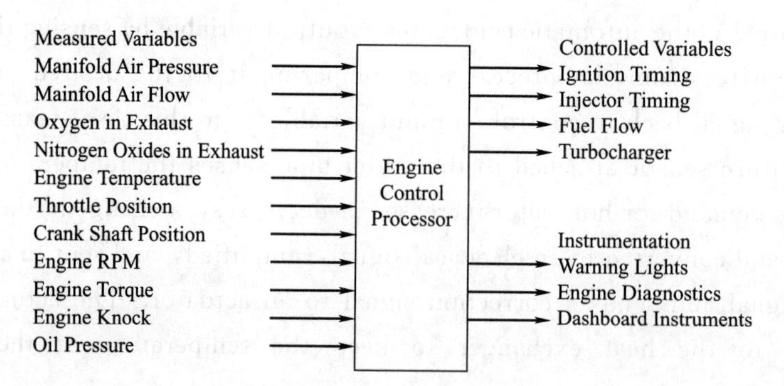

Fig. 20. 5 Automotive engine showing some of the measured and controlled variables

Definitions of the Elements in a Control Loop

Fig. 20. 6 breaks down the individual elements of the blocks in a process-control loop. The measuring element consists of a sensor, a transducer, and a transmitter with its own regulated power supply. The control element has an actuator, a power control circuit, and its own power supply. The controller has a processor with a memory and a summing circuit to compare the set point to the sensed signal so that it can generate an error signal. The processor then uses the error signal to generate a correction signal to control the actuator and the input variable. The definition of these blocks is given as follows:

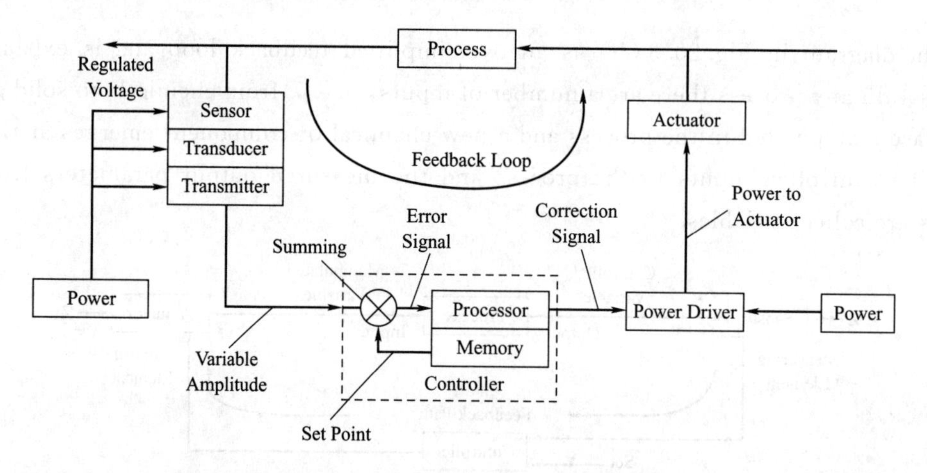

Fig. 20. 6 Block diagram of the elements that make up the feedback path in a process-control loop

Feedback loop is the signal path from the output back to the input to correct for any variation between the output level from the set level. In other words, the output of a process is being continually monitored, the error between the set point and the output parameter is determined, and a correction signal is then sent back to one of the process inputs to correct for changes in the measured output parameter.

Controlled or measured variable is the monitored output variable from a process. The value of the monitored output parameter is normally held within tight given limits.

Manipulated variable is the input variable or parameter to a process that is varied by a control signal from the processor to an actuator. By changing the input variable the value of the measured variable can be controlled.

Set point is the desired value of the output parameter or variable being monitored by a sensor. Any deviation from this value will generate an error signal.

Instrument is the name of any of the various device types for indicating or measuring physical quantities or conditions, performance, position, direction, and the like.

Sensors are devices that can detect physical variables, such as temperature, light intensity, or motion, and have the ability to give a measurable output that varies in relation to the amplitude of the physical variable. The human body has sensors in the fingers that can detect surface roughness, temperature, and force. A thermometer is a good example of a line-of-sight sensor, in that it will give an accurate visual indication of temperature. In other sensors such as a diaphragm pressure sensor, a strain transducer may be required to convert the deformation of the diaphragm into an electrical or pneumatic signal before it can be measured.

Transducers are devices that can change one form of energy to another, e. g. , a resistance thermometer converts temperature into electrical resistance, or a thermocouple converts temperature into voltage. Both of these devices give an output that is proportional to the temperature. Many transducers are grouped under the heading of sensors.

Converters are devices that are used to change the format of a signal without changing the energy form, i. e. , a change from a voltage to a current signal.

Actuators are devices that are used to control an input variable in response to a signal from a controller. A typical actuator will be a flow-control valve that can control the rate of flow of a fluid in proportion to the amplitude of an electrical signal from the controller. Other types of actuators are magnetic relays that turn electrical power on and off. Examples are actuators that control power to the fans and compressor in an air-conditioning system in response to signals from the room temperature sensors.

Controllers are devices that monitor signals from transducers and take the necessary action to keep the process within specified limits according to a predefined program by activating and controlling the necessary actuators.

Programmable logic controllers (PLC) are used in process-control applications, and are microprocessor-based systems. Small systems have the ability to monitor several variables and control several actuators, with the capability of being expanded to monitor 60 or 70 variables and control a corresponding number of actuators, as may be required in a petrochemical refinery. PLCs, which have the ability to use analog or digital input information and output analog or digital control signals, can communicate globally with other controllers, are easily programmed on line or off line, and supply an unprecedented amount of data and information to the operator. Ladder networks are normally used to program the controllers.

An error signal is the difference between the set point and the amplitude of the measured variable.

A correction signal is the signal used to control power to the actuator to set the level of the input variable.

Transmitters are devices used to amplify and format signals so that they are suitable for transmission over long distances with zero or minimal loss of information. The transmitted signal can be in one of the several formats, i. e. , pneumatic, digital, analog voltage, analog current, or as a radio frequency (RF) modulated signal. Digital transmission is preferred in newer systems because the controller is a digital system, and as analog signals can be accurately digitized, digital signals can be transmitted without loss of information. The controller compares the amplitude of the signal from the sensor to a predetermined setpoint, which in Fig. 20. 3(b) is the amplitude of the signal of the hot water sensor. The controller will then send a signal that is proportional to the difference between the reference and the transmitted signal to the actuator telling the actuator to open or close the valve controlling the flow of steam to adjust the temperature of the water to its set value.

Selected from "Fundamentals of Industrial Instrumentation and Process Control", William C. Dunn, The McGraw-Hill Companies, Inc. 2005.

New Words and Expressions

1. HVAC=Heating Ventilation and Air Conditioning 供暖通风与空气调节
2. variable [ˈveriəbl] *a.*; *n.* 变化的，可变的；变量
3. interdependent variable 互依变量
4. obsolete [ˌɔbsəˈliːt] *a.*; *n.*; *vt.* 已过时的；被废弃的事物；淘汰，废弃
5. actuator [ˈæktjueitə] *n.* 执行机构（元件），驱动器，促动器
6. summing circuit 加法［求和］电路
7. variation [ˌveriˈeiʃən] *n.* 变化，变动，演变
8. manipulated variable 操纵量，调节变量
9. line-of-sight 视线
10. magnetic relays 磁继电器
11. unprecedented [ʌnˈpresidentid] *a.* 前所未有的，空前的，没有先例的
12. ladder networks 梯形网络
13. transmitter [trænsˈmitə] *n.* 发送器，发射机
14. modulated signal 调制信号

Unit 21 • The World of Sensors

Sensors are devices that provide an interface between electronic equipment and the physical world. They help electronics to "see", "hear", "smell", "taste", and "touch" . In their interface with the real world, sensors typically convert nonelectrical physical or chemical quantities into electrical signals. This chapter provides a brief introduction to the world of sensors.

Introduction

Microsensors have become an essential element of process control and analytical measurement systems, finding countless applications in, for example, industrial monitoring, factory automation, the automotive industry, transportation, telecommunications, computers and robotics, environmental monitoring, health care, and agriculture; in other words, in almost all spheres of our life. The main driving force behind this progress comes from the evolution in the signal processing. With the development of microprocessors and application-specific integrated circuits (IC), signal processing has become cheap, accurate, and reliable—and it increased the intelligence of electronic equipment①. In the early 1980s a comparison in performance/price ratio between microprocessors and sensors showed that sensors were behind. This stimulated research in the sensor area, and soon the race was on to develop sensor technology and new devices. New products and companies have emerged from this effort, stimulating further advances of microsensors. Application of sensors brings new dimensions to products in the form of convenience, energy savings, and safety. Today, we are witnessing an explosion of sensor applications. Sensors can be found in many products, such as microwave and gas ovens, refrigerators, dishwashers, dryers, carpet cleaners, air conditioners, tape recorders, TV and stereo sets, compact and videodisc players. And this is just a beginning.

Sensor Classification

Sensing the real world requires dealing with physical and chemical quantities that are diverse in nature. From the measurement point of view, all physical and chemical quantities (measurands) can be divided into six signal domains.

① The thermal signal domain: the most common signals are temperature, heat, and heat flow.

② The mechanical signal domain: the most common signals are force, pressure, velocity, acceleration, and position.

③ The chemical signal domain: the signals are the internal quantities of the matter such as concentration of a certain material, composition, or reaction rate.

④ The magnetic signal domain: the most common signals are magnetic field intensity, flux density, and magnetization.

⑤ The radiant signal domain: the signals are quantities of the electromagnetic waves such as intensity, wavelength, polarization, and phase.

⑥ The electrical signal domain: the most common signals are voltage, current, and charge.

As mentioned, sensors convert nonelectrical physical or chemical quantities into electrical signals. It should be also noted that the principle of operation of a particular sensor is dependent on the type of physical quantity it is designed to sense. Therefore, it is no surprise that a general classification of sensors follows the classification of physical quantities. Accordingly, sensors are classified as thermal, mechanical, chemical, magnetic, and radiant.

There is also a classification of sensors based on whether they use an auxiliary energy source or not. Sensors that generate an electrical output signal without an auxiliary energy source are called self-generating or passive. An example of this type of sensors is a thermocouple. Sensors that generate an electrical output signal with the help of an auxiliary energy source are called modulating or active.

Sensor Parameters

Performance of sensors, like other electronic devices, is described by parameters. The following section briefly describes the most common sensor parameters.

① Absolute sensitivity is the ratio of the change of the output signal to the change of the measurand (physical or chemical quantity).

② Relative sensitivity is the ratio of a change of the output signal to a change in the measurand normalized by the value of the output signal when the measurand is 0.

③ Cross sensitivity is the change of the output signal caused by more than one measurand.

④ Direction dependent sensitivity is a dependence of sensitivity on the angle between the measurand and the sensor②.

⑤ Resolution is the smallest detectable change in the measurand that can cause a change of the output signal.

⑥ Accuracy is the ratio of the maximum error of the output signal to the full-scale output signal expressed in a percentage.

⑦ Linearity error is the maximum deviation of the calibration curve of the output signal from the best fitted straight line that describes the output signal③.

⑧ Hysteresis is a lack of the sensor's capability to show the same output signal at a given value of measurand regardless of the direction of the change in the measurand.

⑨ Offset is the output signal of the sensor when the measurand is 0.

⑩ Noise is the random output signal not related to the measurand.

⑪ Cutoff frequency is the frequency at which the output signal of the sensor drops to 70.7% of its maximum.

⑫ Dynamic range is the span between the two values of the measurand (maximum and minimum) that can be measured by sensor.

⑬ Operating temperature range is the range of temperature over which the output signal of the sensor remains within the specified error.

It should be pointed out that in addition to these common parameters, other parameters are often used to describe other unique properties of sensors.

Selected from "Sensor Technology & Devices", Ljubisa Ristic, Artech House, Inc., 1994.

New Words and Expressions

1. sensor ['sensə] *n*. 传感器，检测器
2. microsensor [maikrəu'sensə] *n*. 微型传感器
3. nonelectrical [nɔni'lektrik(ə)l] *a*. 不用电的，非电的
4. microprocessor [maikrəu'prəusesə (r)] *n*. 微处理器，微信息处理机，微型计算机
5. measurand ['meʒərənd] *n*. 被测的物理量，测量变量
6. flux density 通量密度
7. polarization [ˌpəulərai'zeiʃən] *n*. 极化，偏振
8. modulating ['mɔdjuleitiŋ] *a*. 调制的
9. calibration [ˌkæli'breiʃən] *n*. 校准
10. hysteresis [ˌhistə'riːsis] *n*. 滞后［现象，效应］，磁滞现象
11. cutoff frequency 截止频率

Notes

① 参考译文：随着微型计算机和特定用途的集成电路的发展，信号处理变得价廉、精确而且可靠——它提高了电子设备的智能化。

② 参考译文：方向相关灵敏度是与被测变量和传感器之间的角度相关的灵敏度。

the dependence of *M* on*N*　*M* 对 *N* 的依存关系［曲线关系］

③ 参考译文：线性误差是输出信号的校准曲线与最适合描述输出信号的直线之间的最大偏差。

the deviation of *M* from *N*　*M* 与 *N* 不一致［有偏差］

Exercises

1. After reading the text above, write a summary of it.
2. Answer the following questions according to the text.
   ① What are the sensors?
   ② Please list some examples of sensor applications.
   ③ What is the general classification of sensors?
   ④ What is the absolute sensitivity of the sensor?
3. Translate the 2nd paragraph of the text into Chinese.
4. Put the following into Chinese by reference to the text.
   integrated circuit　　signal processing　　performance/price ratio　　stimulate

magnetic field intensity　magnetization　thermocouple　cross sensitivity
auxiliary energy source　resolution

5. Put the following into English.
微型传感器　通量密度　电磁波　线性误差
相对灵敏度　传感器参数偏移

6. Translate the following sentences into English.
① 微型传感器是过程控制和分析测量系统必不可少的部件，在我们生活的各方面都有无数的应用。
② 绝对灵敏度是输出信号的变化与被测变量变化之比。

Reading Material 21

Signal Conversion

Conversion Systems

Conversion systems can be used to transform analog signals into digital form for digital-computer processing. In some applications, such as digital control systems, digital signals are also transformed to analog signals for controlling the continuous plant. Some analog signal processing is usually employed prior to analog-to-digital (A/D) conversion (an anti-aliasing filter) or after the digital-to-analog (D/A) conversion (a smoothing filter). Fig. 21. 1 represents an overall block diagram for a general conversion system. A transducer transforms a physical parameter such as pressure, temperature, velocity, or position into an analog signal (current, voltage, frequency, etc.). Further analog processing is accomplished by using filtering (amplification, spectrum changing, noise depression, etc.). The analog signal is then converted into digital form (binary-coded) for transfer into the digital processor. After appropriate transfer function digital processing, digital signals are transferred to a memory device (register, computer memory, etc.) for storage and then conver-

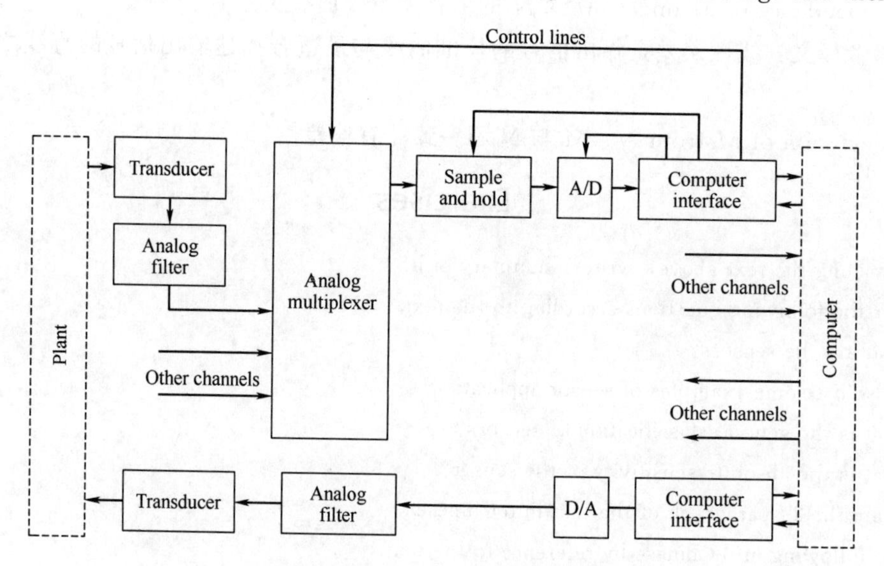

Fig. 21. 1　General conversion-system organization for control

ted to an analog signal for control purposes.

In general, various input and output ports (interfaces) exist on the digital processor (computer) such that numerous channels can be connected. A single port can transfer a number of channels through the use of a multiple sharing device known as a multiplexer. Each input channel is connected to the multiplexer output for a specified period of time as controlled by switches. The circuits which follow the multiplexer are thus time-shared. A sample-and-hold device samples the multiplexer output and holds the voltage level (or equivalent) such that the A/D converter has an appropriate period of time to complete its operation.

The sequencing of the A/D process is controlled by various signals that are generated by the computer interface under software and/or hardware control. The D/A process is usually simpler to control since the output transfer is synchronized to the computer cycle period, and thus data are transferred directly to the output without any interface "handshaking".

General Analog-to-digital (A/D) Conversion Structures

An analog-to-digital (A/D) converter or ADC transforms an analog signal to a q-bit computer word generally through the use of the comparison operation. The first step of the operation compares an unknown input signal (voltage, current, etc.) with a known value and determines which is larger. This known value, V_k, is varied according to

$$V_k = V_{ref} \sum_{i=1}^{q} b_i 2^{-i}$$

which is one of 2^q possible words nearest to the unknown input value. The next step of the conversion process can be accomplished by using a variety of techniques and organizations to minimize the error between these values. Organizational parameters that define most types of A/D conversion systems involve the following seven elements:

Basic reference technique;

Data transfer (serial-parallel) organization;

Information flow structure;

Timing structure;

Comparator operation;

Digital coding of binary information;

Scaling of binary information.

Specific Analog-to-digital Conversion Systems

Analog-to-digital converter structures reflect various organizational design parameters as mentioned in the last section. Each affects overall converter performance, efficiency, and economy in terms of a specific application. The required degree of accuracy, e. g. , depends upon the frequency of the analog input signal, the conversion time, and associated A/D electronic errors. The conversion delay between the A/D START command and the generation of the desired binary value is different for various A/D structures. The meaning of specific terms and parameters must be understood in selecting commercial devices. Beyond those mentioned in the previous section, the following terms are defined.

Accuracy: exactness of the digital value to the true analog value. Accuracy is constrained by the converter noise, aperture time, quantization (1/2 LSB), and the electronic switching process (linearity, gain, temperature, etc.).

Aperture time: the time interval of uncertainty or the "window" in which the analog value occurred. The specific aperture time required in a given application is directly related to the acceptable error between an analog input magnitude within the aperture window and the value of the associated digitized value. This type of error occurs because the input analog signal may change in magnitude during the digitizing process, and thus the digitized value is associated with only one analog value that occurred in the window.

Conversion time: time required to transform the analog input into a binary value after a converter START command is given. For some converter organizations, this time is constant; for others, it is a function of the current input amplitude, i. e., the time required to acquire the analog signal in a stepwise fashion.

Sampling time: *T*, as defined previously. In reality, *T* is the time between START commands to the converter. *T* is usually much, much greater than the conversion time because of the theoretical model (impulse sampling) used in the determination of the digital compensator.

The fundamental component of most A/D converters is an analog signal comparator that has a positive logic output of "1" if the analog input signal is greater than an analog reference voltage. If the input value is lower than the reference voltage, the comparator output is defined to be a logical "0". To generate a digitized value of the input, additional logic is added to the comparator circuitry to manipulate the reference voltage and to minimize the error between the input signal (assumed constant) and the reference voltage. The digitized and encoded value of the manipulated reference is the desired value.

Selected from "Digital Control Systems—Theory, Hardware, Software" (Second edition), Constantine H. Houpis and Gary B. Lamont, McGraw-Hill, Inc., 1992.

New Words and Expressions

1. antialiasing filter 去假频滤波器
2. smoothing filter 平滑滤波器
3. spectrum changing 频谱改变
4. binary-coded *a*. 二进制编码的
5. multiplexer [ˈmʌltiˌpleksə] *n*. 多路调制［转换］器，多路开关选择器
6. time-shared *a*. 分时的
7. sample-and-hold 样品保持
8. handshake [ˈhændʃeik] *n*.; *v*. 符号［信息］变换，握手信号
9. bit [bit] *n*. 比特（二进位数），位，存储信息容量单位
10. serial-parallel 串行-并行
11. comparator [ˈkɔmpəreitə] *n*. 比较器［块，装置，电路］

12. linearity [ˌliniˈærəti] *n*. 直线性，线性
13. gain [gein] *n*. 增益（系数），放大（系数，率），增量
14. stepwise [ˈstepwaiz] *a*.; *ad*. 逐步（的），逐渐（的），分段的，阶式的
15. circuitry [ˈsəːkitri] *n*. 电路系统，（整机）电［网］路

Unit 22 • Introduction to Vibration

Of all the parameters that can be measured in industry today, none contains as much information as the vibration signature. Even though proper maintenance of industrial equipment may require the measuring or monitoring of other parameters such as: temperature, pressure, flow, voltage, current, horsepower and torque, the fact remains that the vibration signature contains more information about the machine's health and operating characteristics than any other parameter.

That is not to say that we are presently able to extract all of the information contained in that vibration signature as we attempt to analyze a piece of equipment. However, our ability to extract and analyze information from vibration data has grown dramatically in the last 15 to 20 years.

Early Methods of Vibration Analysis

No more than a generation ago, the quality of operating equipment from a vibration standpoint was determined by using a coin on edge or feeling the bearings of the machine. If the coin remained vertical while the machine was running or the bearings felt good to the inspector, the vibration was assumed to be acceptable. On the other hand, if the coin fell over or the bearings felt bad, an attempt was made to balance some part of the machine.

Plant personnel quite often concluded that a machine vibrated because it was out of balance; other causes of machine vibration (such as misalignment, bent shaft, hydraulic instability, or bearing failure) were rarely suspected①. Large machine tool manufacturers would ship what was considered to be high—quality machinery, where the only test for vibration was how the machine "felt" to the quality control inspector or manager.

Rotors were balanced by using chalk or soapstone held lightly against the shaft. A mechanic would look at the arc of the chalk mark and estimate the magnitude and location of the imbalance. Mechanics would also hold a pointer close to a shaft and, by eye, judge the magnitude of the vibration by watching the varying gap between the pointer and the shaft.

As frequencies increased, it became much more difficult to estimate the amount of vibration. It was not uncommon to mount dial indicators against a bearing housing or a shaft, and to use the dial indicator to read the vibration amplitude. As shaft frequencies increased the mass integration of the dial indicator made it impossible to read the displacement accurately, if at all. That is, for a given input energy, a mass has greater displacement at lower frequencies than at high frequencies. So the mass of the dial indicator moves less at high frequencies than at low frequencies, and is therefore less sensitive at these higher frequencies.

A light beam vibrometer was developed which had very low mass integration. This device could read vibration amplitudes at frequencies as high as 2,000 CPM (cycles per minute) without degradation, and as high as 4,000 CPM with some loss of reliability and accuracy.

Uses of Vibration Analysis

The collection and reduction of vibration data allows you to perform three important analysis techniques: (1) Certification; (2) Potential fault analysis; (3) Diagnostic analysis.

Certification

Vibration data is taken on new and/or rebuilt machinery to ensure that the machinery is operating within acceptable vibration tolerances. If excessive vibration is discovered, the cause can be remedied before the machinery is put into service.

Potential Fault Analysis

Also called "predictive" maintenance, potential fault analysis allows you to determine the health of machine components. A good assessment of machine health allows you to predict what component is most likely to fail first, and approximately how long that component will continue to operate before fault conditions begin to occur②. This information allows you to anticipate and schedule machine overhauls, thereby minimizing the impact of process shutdowns and avoiding damage that results from running a machine until a catastrophic failure occurs.

In potential fault analysis, information about the machine is used to determine the number and location of data collection points. Machine information includes:

① the number, type, and location of each bearing;

② the type and location of couplings;

③ machine components (turbine, pump, generator, etc.).

These topics are discussed in detail in subsequent chapters.

Mathematical techniques are then used to calculate approximate fault frequencies for each system component. If possible, the phase relationships of component faults can also be determined.

Possible causes for each fault frequency can also be determined, and each fault frequency is prioritized. For example, probability of each fault is determined, the seriousness of that fault is assessed including the cost of down time and repair, and appropriate collection intervals are scheduled.

The vibration data is analyzed to determine if any of the fault frequencies are present and, if so, their severity. Vibration tolerances, set in conjunction with manufacturer's specifications, are used to determine severity. Proximity to an alarm condition indicates the most likely fault and the trend allows you to approximate when it will occur.

A carefully maintained potential fault analysis program, coupled with a program of machine maintenance, ensures long, trouble—free operation of plant machinery as well as better product quality.

Diagnostic Analysis

Whereas potential fault analysis is more proactive, diagnostic analysis is reactive: it is only performed once a problem has occurred. Often times the problem will be so severe that equipment must be shut down for safety reasons.

Diagnosis begins with the same steps as potential fault analysis. Machine data is used to determine the number and location of data collection points，and fault frequencies and phase relationships are calculated.

Once the vibration data is collected，a comparison is made between the calculated fault frequencies and their harmonics and the actual frequencies that dominate the vibration signature. The cause of each fault frequency is determined，and corrective measures are taken to correct the problem. Follow-up readings are taken to ensure that the problem was indeed solved and that the machine is operating within tolerance.

If a potential fault analysis program is not already in place，then the best time to begin the program is immediately after a diagnostic analysis. The machine data，fault frequencies，and possible causes have been determined. An inspection of the machine will have been completed during the diagnosis to ensure that no other causes of vibration are present，such as：loose bolts，cracked weldments，grouting turned to powder，ash accumulation on fan blades.

Now that a thorough inspection of the machine has taken place，the machine is operating smoothly and the current state of the machine is fresh in the operator's and maintenance manager's mind. A potential fault analysis program should be implemented now to ensure that the machine remains in good health.

Selected from "The simplified handbook of vibration analysis"，Authur R. Crawford，Computational System Incorporated，1992.

New Words and Expressions

1. vibration signature 振动波形，振动特征
2. monitor [ˈmɔnitə] *n.* 监测
3. balance [ˈbæləns] *n.* 平衡，对称
4. misalignment [ˈmisəlaimənt] *n.* 不对中，不同心度，不平行度
5. hydraulic instability 液压不稳定，水力不稳定
6. amplitude [ˈæmplitjuːd] *n.* （振，波）幅，幅度
7. predictive maintenance 预测性维修
8. harmonics [haːˈmɔniks] *n.* 谐（调和）函数，谐波（频，音）

Notes

① 参考译文：工厂技术人员常将机器振动原因归结为不平衡，而很少怀疑其它原因（如：不对中、轴弯曲、液压不稳定或轴承失效等）。

② 参考译文：机器健康状况的精密诊断允许你能预测哪个零件会首先失效，零件失效前还能运行多长时间。

Exercises

1. After reading the text above，write a summary of it.
2. Answer the following questions according to the text.

① What is the predictive maintenance?

② Please list measuring or monitoring parameters of the industrial equipment.

③ Please list the main causes of machine vibration.

④ What are concluded in the machine information?

3. Translate the 6th paragraph into Chinese.

4. Put the following into Chinese by reference to the text.

misalignment　　bent shaft　　hydraulic instability　　bearing failure　　loose bolts

5. Translate the following sentences into English.

① 新的或改造过的机器要进行振动测试，以保证机器在允许的范围内运行。

② 一旦得到振动数据，就要对计算的故障频率、谐振频率和振动波形中的主频率进行比较。

Reading Material 22

Vibration Analysis and Fault Diagnosis

All mechanical systems are forced to vibrate by one or more forcing functions. In some cases this force may be small, and natural and/or designed sources of damping within the system keep the vibration within acceptable levels.

However, a forcing function may cause a system to vibrate excessively. This vibration can result in bearing failure, fatigue and cracking in machine components, and unacceptable damage to the product being produced. Left uncorrected, excessive vibration can destroy entire structures and result in serious injury or death.

In essence, a vibration problem is solved by identifying the forcing function (s) and eliminating or minimizing their effect. While this is not always as simple as it may sound, a general step-by-step procedure may be followed during analysis of a vibration problem. This section provides an outline of this procedure and discusses a variety of analysis techniques that can aid in vibration problem solving.

Outline to Vibration Problem Solving

The analysis of a vibration problem can be broken into seven steps:

① Identify the problem;

② Gather information about the vibrating system;

③ Determine the possible forcing function (s);

④ Determine where to take data and what equipment to use;

⑤ Take vibration data;

⑥ Analyze vibration data;

⑦ Make recommendations.

Identify the Problem

Not all machine problems attributed to vibration are, in fact, vibration. Noise is quite often mistaken as vibration. When approaching a vibration problem, the first step is to determine if the problem is actually vibration; if the problem is noise, then the vibration analysis techniques discussed here will not be effective.

Sometimes noise and vibration are combined. For example, a piece of metal may be loose inside a paper roll. As the roll turns the metal bangs around inside, causing noise and

vibration.

Gather Information about the Vibrating System

The next step in the solution of a vibration problem is to gather as much information as possible about the machine you are going to analyze. There are three essential steps in this information gathering process: ①Sketch the machine and collect all component information (bearings, belts, gears, etc.); ②Gather the available maintenance history; ③Obtain the machine operator's input.

Sketch the Machine

A sketch of the machine allows you to determine the best data collection points for your analysis, and aids you in determining potential sources of vibration.

Gather the Available Maintenance History

The maintenance history of the machine often provides valuable clues to the cause of vibration. Among the things you should determine are:

—What was the last thing done to the machine? If, while researching the maintenance history of a vibrating machine you discover that bearings have been replaced several times due to premature failure, then you can suspect: poor installation techniques, inadequate lubrication procedures, bearing misalignment, bearing load, operating frequency, shaft misalignment, electrical discharge through the bearing, as the cause of vibration.

—What is the maintenance history of the machine? Has it ever been damaged? Have major components (bearings, gears, couplings, belts, rotors, etc.) been replaced or repaired? Has the suspect component had a history of problems?

—How thorough has general maintenance been? Has lubricant been maintained and changed regularly?

—Bearings: How often are clearances checked, and are they still in tolerance? Have any bearings been changed? If so, was the change due to normal wear or premature failure? Do the bearings need to be lubricated? If so, have they been lubricated regularly and properly? Over-lubrication is as serious a problem as under-lubrication. You should also determine which bearings are fixed and which are floating. There should only be one fixed bearing on a shaft. More than one fixed bearing on a single shaft can lead to early bearing failure.

—Shafts: What's the shaft material, and has any work been done on the shaft? If the shaft is made of a high-stress material like cold-rolled steel it will stress relieve.

—Alignment: How were the shaft aligned, and to what tolerance? If there are couplings on the shaft, is the alignment tolerance within the tolerance of the coupling? Was allowance made for thermal growth, and was the thermal growth checked? How often is the alignment checked?

—Check to see if there has been any previous work done on the machine to correct vibration. If so, compare the type of vibration readings taken then to what you have taken now. Similar readings could indicate that a correction weight that was previously installed has come off or that a previous problem is developing again.

Obtain the Machine Operator's Input

The machine operator is a good source of information on the characteristics and severity of the machine's vibration. Operators develop a "feel" for a machine when they work with it every day, and are often aware of developing problems before they become serious.

Determine the Possible Forcing Function (s)

The determination of the forcing functions (the cause, or causes, of the machine's vibration) that are present in the system is the goal of vibration analysis. Once the forcing function has been identified, intelligent decisions can be made for correction of the problem.

The most common forcing functions include: rotor frequencies, bearing fault frequencies, rubs, belt frequencies, gear mesh frequencies, blade pass frequencies, harmonics of any of these frequencies, resonant frequencies.

Determine Where to Take Data and What Equipment to Use

Once you make a sketch of the machine and determine the possible forcing functions, you must choose the data collection locations and the transducer and instrumentation combinations best suited to solve the problem.

You must collect data in the correct plane in order to properly diagnose the vibration. There are two planes for collecting data on a rotor bearing: axial and radial. Readings in the radial plane are taken in both vertical and horizontal orientations. The axial plane is along the center line of the rotor; that is, the shaft's center line lies in the axial plane. Some forcing functions, such as angular misalignment or a bent shaft, have their greatest effect in the axial plane. The radial plane is perpendicular to the axial plane; the radii of the shaft lie in the radial plane.

Forcing functions such as imbalance, rotor rub, or loose bearings dominate the radial plane. There are two radial plane readings that are usually taken: vertical and horizontal. Enough readings should be taken at each measuring point so that the rocking and translational modes of that point in space can be established.

Finally, the vibration data should be repeatable; you should be sure that you would get the same readings if you shut the machine down and started it up again.

Take Vibration Data

Vibration data from a transducer must be converted into a meaningful form for proper diagnosis. There are two types of displays that we will consider: time domain and frequency domain. A time domain display shows how the amplitude of the signal varied over time. Since the amplitude of the transducer's output voltage is proportional to the amplitude of vibration, this display shows the vibration amplitude over time.

Time domain displays are typically sine waves that have undergone harmonic distortion. By using a Fast Fourier Transform (FFT) analyzer, the time domain waveform can be broken down into its pure fundamental and harmonic sine waves.

A frequency domain signal is a plot of the signal amplitude on the Y axis versus frequency on the X axis. Called a frequency signature in the vibration industry, a frequency domain

signal shows the frequencies and amplitudes of periodic components in the signal.

Analyze Vibration Data

Using the phase and frequency information, you can now make a good estimate as to the forcing function (s) present in your system. You may find quite often that more than one forcing function is contributing to the overall system vibration. If this is the case, you should always treat the most severe problem first, then tackle the next problem; always solve one problem at a time.

Make Recommendations

You should determine the probable cause (or causes) of the vibration that you have measured.

Summary

Vibration problems are solved by identifying the forcing function (s) and eliminating or minimizing their effect. A general step-by-step procedure may be followed during analysis of a vibration problem.

Selected from "The simplified handbook of vibration analysis", Authur R. Crawford, Computational System Incorporated, 1992.

New Words and Expressions

1. damping [ˌdæmpiŋ] *a*. 阻尼的，减振的，衰减的
2. electrical conductor 导电体
3. sketch [skætʃ] *n*.; *v*. 草图，示意图；画草图，画示意图
4. coupling [ˈkʌpliŋ] *n*. 偶合，连接，轴接，联轴器（节）
5. cold-rolled [kəuld rəuld] *a*. 冷轧的
6. gear mesh 齿轮啮合
7. resonant [ˈrezənənt] *a*. 共振的，引起共鸣的

PART Ⅲ

COMPUTERIZED MANUFACTURING TECHNOLOGIES

Unit 23 ● What is "Mechatronics"?

"Mechatronics" is a term coined by the Japanese to describe the integration of mechanical and electronic engineering. The concept may seem to be anything but new, since we can all look around us and see a myriad of products that utilize both mechanical and electronic disciplines. Mechatronics, however, specially refers to a multidisciplined, integrated approach to product and manufacturing system design. It represents the next generation of machines, robots, and smart mechanisms necessary for carrying out work in a variety of environments—primarily, factory automation, office automation, and home automation as shown in Fig. 23. 1.

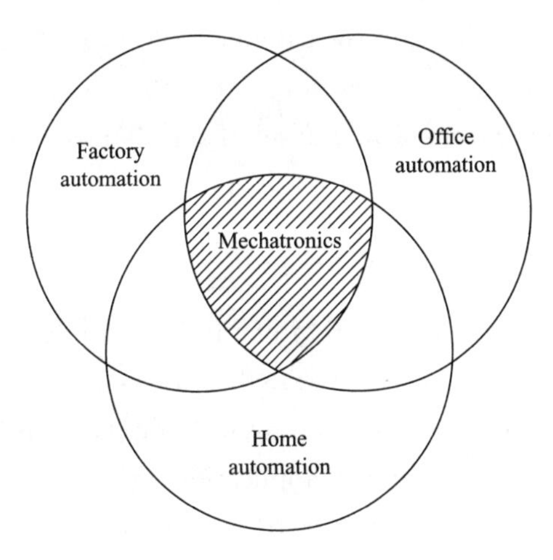

Fig. 23. 1 Mechatronic environments

By both implication and application, mechatronics represents a new level of integration for advanced manufacturing technology and processes. The intent is to force a multidisciplinary approach to these systems as well as to reemphasize the role of process understanding and control. This mechatronic approach is currently speeding up the already-rapid Japanese process for transforming ideas into products.

Currently, mechatronics describes the Japanese practice of using fully integrated teams of product designers, manufacturing, purchasing, and marketing personnel acting in

concert with each other to design both the product and the manufacturing system①.

The Japanese recognized that the future in production innovation would belong to those who learned how to optimize the marriage between electronic and mechanical systems. They realized, in particular, that the need for this optimization would be most intense in application of advanced manufacturing and production systems where artificial intelligence, expert systems, smart robots, and advanced manufacturing technology systems would create the next generation of tools to be used in the factory of the future②.

From the very beginnings of recorded time, mechanical systems have found their way into every aspect of our society. Our simplest mechanisms, such as gears, pulleys, springs, and wheels, have provided the basis for our tools. Our electronics technology, on the other hand, is completely twentieth-century, all of it created within the past 75 years.

Until now, electronics were included to enhance mechanical systems' performance, but the emphasis remained on the mechanical product. There had never been any master plan on how the integration would be done. In the past, it had been done on a case-by-case basis. More recently, however, because of the overwhelming advances in the world of electronics and its capability to physically simplify mechanical configurations, the technical community began to reassess the marriage between these two disciplines.

The most obvious trend in the direction of mechatronic innovation can be observed in the automobile industry. There was a time when a car was primarily a mechanical marvel with a few electronic appendages.

First came the starter motor, and then the generator, each making the original product a bit better than it was before. Then came solid-state electronics, and suddenly the mechanical marvel became an electro-mechanical marvel. Today's machine is controlled by microprocessors, built by robots, and fault-analyzed by a computer connected to its "external interface connector"③. Automotive mechanical engineers are no longer the masters of their creations.

The process that describes the evolution of the automobile is somewhat typical of other products in our society. Electronics has repeatedly improved the performance of mechanical systems, but that innovation has been more by serendipity than by design. And that is the essence of mechatronics—the preplanned application of, and the efficient integration of, mechanical and electronics technology to create an optimum product.

A recent U. S. Department of Commerce report entitled "JTECH Panel report on Mechatronics in Japan" compared U. S. and Japanese research and development trends in specific areas of mechatronics technology. Except for a few areas, the technology necessary to accomplish the development of the next generation of systems embodying the principles of mechatronics is fully within the technological reach of the Japanese.

Comparisons were made in three categories: basic research, advanced development, and product implementation④. Except for machine vision and software, Japanese basic research was comparable to the United States, with the Japanese closing in fast on machine vision system technology. Japanese artificial intelligence research is falling behind, primarily

because the Japanese do not consider it an essential ingredient of their future systems, they appear capable of closing even that gap, if required. In the advanced development and product implementation areas, Japan is equal to or better than the United States, and is continuing to pull ahead at this time.

The Department of Commerce report concluded that Japan is maintaining its position and is in some cases gaining ground over the United States in the application of mechatronics. Their progress in mechatronics is important because it addresses the very means for the next generation of data-driven advanced design and manufacturing technology. In fact, the Department of Commerce report concludes that this has created a regenerative effect on Japan's manufacturing industries.

To close the gap, we will need to go much further than creating new tools. If we accept the fact that mechanical systems optimally coupled with electronics components will be the wave of the future, then we must also understand that the ripple effect will be felt all the way back to the university, where we now keep the two disciplines of mechanics and electronics separated and allow them to meet only in occasional overview sessions. New curricula must be created for a new hybrid engineer—a mechatronics engineer. Only then can we be assured that future generations of product designers and manufacturing engineers will fully seek excellence in these new techniques.

We need to rethink our present-day approach of separating our engineering staffs both from each other and from the production engineers. Living together and communicating individual knowledge will create a new synergistic effect on products. Maximum interaction will be the key to optimum designs and new product development.

The definition of mechatronics is much more significant than its combined words imply. It can physically turn engineering and manufacturing upside down. It will change the way we design and produce the next generation of high technology products. The nation that fully implements the rudiments of mechatronics and vigorously pursues it will lead the world to a new generation of technology innovation with all its profound implications⑤.

Selected from "Mechatronics: Japan's Newest Threat", V. Daniel Hunt, Chapman and Hall, 1988.

New Words and Expressions

1. mechatronics [mi'kætrɔniks] *n*. 机械电子学，机电一体化
2. integration [ˌinti'greiʃən] *n*. 综合，结合，一体化，集成
3. myriad ['miriəd] *a*. 无数，无数的人 [或物]
4. multidisciplined [ˌmʌlti'disiplind] *a*. 多学科的
5. reemphasize [ri'emfəsaiz] *v*. 反复强调
6. innovation [ˌinəu'veiʃən] *n*. 改革，革新
7. reassess [ˌriːə'ses] *v*. 重估，再评价
8. marvel ['mɑːvl] *n*. 奇异事物，奇迹
9. appendage [ə'pendidʒ] *n*. 附加物，附件

10. serendipity [serən'dipəti] *n*. 运气，善于发掘新奇事物的才能
11. essence ['esns] *n*. 实质，本质
12. implementation [implimen'teiʃən] *v*. 实施，实行，履行
13. be coupled with 和……联合，结合
14. ripple ['ripl] *n*. 波纹，涟漪
15. hybrid ['haibrid] *a*. 混合的
16. synergistic ['sinədʒistik] *a*. 协同的，协作的，叠加的，复合的

Notes

① 参考译文：目前机电一体化描述的是，利用产品设计人员与制造、采购、营销等人员全部集成为团队的日本实践，这类团队在设计产品与制造系统中相互协同行动。

acting in concert with（同……一致行动）是现在分词短语修饰 personnel。

② 参考译文：他们认识到在用人工智能、专家系统、灵巧机器人和先进制造技术来制造出将来工厂要使用到的新一代工具的制造和生产系统，尤其需要这种机电一体化的结合。

need for …：……的需要；artificial intelligence：人工智能；expert system：专家系统。

③ 参考译文：今天的机器都是有微处理器控制，机器人制造，由和外接口连接器相连的计算机进行故障分析。

全句有三个并列句组成，主语都是 Today's machine，谓语分别是 is controlled，(is) built，(is) fault-analyzed。

④ 参考译文：在三个方面做对比：基础研究，预先研制，产品实现。

⑤ 参考译文：彻底贯彻机电一体化的基本原理，并强力践行机电一体化的国家，将引领一场具有机电一体化全部深邃蕴涵的新一代技术创新。

Exercises

1. After reading the text above, write a summary of it.
2. Answer the following questions according to the text.
   ① What is the intent of mechatronics?
   ② Please describe the evolution of the automobile.
   ③ What are the three categories of comparisons in mechatronics between U. S. and Japan?
   ④ Why must the new curricula be created for a mechatronics engineer?
3. Translate the 7th and 10th paragraphs of the text into Chinese.
4. Put the following into Chinese by reference to the text.
   innovation　microprocessor　machine vision　appendage　discipline　automobile
5. Put the following into English.
   机器人　结构（外形）　接口　发电机　人工智能　专家系统
6. Translate the following sentences into English.
   ① 机电一体化概念并非新概念，因为看看我们周围有数不清的机械电子产品。
   ② 除了机器视觉和软件外，机电一体化基础研究日本和美国差不多。

Reading Material 23

Benefits of Mechatronics

Mechatronics may sound like utopia to many product and manufacturing managers because it is often presented as the solution to nearly all of the problems in manufacturing. The primary benefits of mechatronics, with an emphasis on advanced manufacturing technology and factory automation, are summarized below.

High Capital Equipment Utilization

Typically, the throughput for a set of machines in a mechatronics system will be up to three times that for the same machines in a stand-alone job shop environment. The mechatronic system achieves high efficiency by having the computer schedule every part to a machine as soon as it is free, simultaneously moving the part on the automated material handling system and downloading the appropriate computer program to the machine. In addition, the part arrives at a machine already fixtured on a pallet (this is done at a separate work station) so that the machine does not have to wait while the part is set up.

Reduced Capital Equipment Costs

The high utilization of equipment results in the need for fewer machines in the mechatronic system to do the same work load as in a conventional system. Reductions of 3 : 1 are common when replacing machining centers in a job-shop situation with a mechatronic system.

Reduced Direct Labor Costs

Since each machine is completely under computer control, full-time oversight is not required. Direct labor can be reduced to the less skilled personnel who fixture and defixture the parts at the work station, and a machinist to oversee or repair the work stations, plus the system supervisor. While the fixturing personnel in mechatronic environments require less advanced skills than corresponding workers in conventional factories, labor cost reduction is somewhat offset by the need for computing and other skills which may not be required in traditional workplaces.

Reduced Work-in-Process Inventory and Lead Time

The reduction of work-in-process in a mechatronic system is quite dramatic when compared to a job-shop environment. Reductions of 80 percent have been reported at some installations and may be attributed to a variety of factors which reduce the time a part waits for metal-cutting operations. These factors include concentration of all the equipment required to produce part into a small area; reduction in the number of fixtures required; reduction in the number of machines a part must travel through because processes are combined in work cells; and efficient computer scheduling of parts batched into and within the mechatronic system.

Responsiveness to Changing Production Requirements

A mechatronic system has the inherent flexibility to manufacture different products as

the demands of the marketplace change or as engineering design changes are introduced. Furthermore, required spare part production can be mixed into regular runs without significantly disrupting the normal mechatronic system production activities.

Ability to Maintain Production

Many mechatronic systems are designed to degrade gracefully when one or more machines fail. This is accomplished by incorporating redundant machining capability and a material handling system that allows failed machines to be bypassed. Thus, throughput is maintained at a reduced rate.

High Product Quality

A sometimes-overlooked advantage of a mechatronic system, especially when compared to machines that have not been federated into a cooperative system, is improved product quality. The basic integration of product design characteristics with production capability, the high level of automation, the reduction in the number of fixtures and the number of machines visited, better designed permanent fixtures, and greater attention to part/machine alignment all result in good individual part quality and excellent consistency from one workpiece to another, further resulting in greatly reduced costs of rework.

Operational Flexibility

Operational flexibility offers a significant increment of enhanced productivity. In some facilities, mechatronic systems can run virtually unattended during the second and third shifts. This nearly "unmanned" mode of operation is currently the exception rather than the rule. It should, however, become increasingly common as better sensors and computer controls are developed to detect and handle unanticipated problems such as tool breakages and part-flow jams. In this operational mode, inspection, fixturing, and maintenance can be performed during the first shift.

Capacity Flexibility

With correct planning for available floor space, a mechatronics system can be designed for low production volumes initially; as demand increases, new machines can be added easily to provide the extra capacity required.

Selected from "Mechatronics: Japan's newest Threat", V. Daniel Hunt, Chapman and Hall, 1988.

New Words and Expressions

1. Utopia [juːˈtəupiə] *n.* 乌托邦，理想的完美境界
2. simultaneously [ˌsiməlˈteiniəsli] *ad.* 同时，同时发生
3. inherent [inˈhiərənt] *a.* 固有的，先天的，内在的
4. federate [ˈfedəreit] *v.* [使] 联合，联盟
5. alignment [əˈlainmənt] *n.* 对齐，校准，调整
6. increment [ˈinkrimənt] *n.* 增量，递增，增值

Unit 24 • Computerized Numerical Control (1)

Numerical Control

This century has seen machinery become more automated, thereby eliminating machine-operator intervention in the manufacturing process. Yet, as automation has increased, machines have become inherently more specialized. A highly automated machine that may turn out 20,000 components per day, will generally be able to produce only a limited class of components. Until recently, prototypes and low volume components were produced by manually operated conventional machine tools.

With the advent of new hard-to-machine materials and requirements for tolerances of a precision that can approach one part in ten thousand, the best human operators have reached the limited class of their ability①. These requirements, together with a need for component flexibility, have lead to a form of automatic machine control known by the generic name *numerical control* (frequently abbreviated NC in this text and in engineering literature generally).

The history of NC began in the late 1940s, when John T. Parsons proposed that a method of automatic machine control be developed which would guide a milling cutter to generate a smooth curve. As he conceived it, the coordinate points would be coded onto punched cards fed to a machine controller which would cause a modified milling machine to move in small incremental steps to achieve the desired path②. In 1949 the U. S. Air Force commissioned the Servomechanisms Laboratory at the Massachusetts Institute of Technology to develop a workable NC system based on Parsons' concept③.

Scientists and engineers at M. I. T. selected perforated paper tape as the communication medium and initially built a two-axis point-to-point system which positioned the drilling head over the coordinate. Later, a more sophisticated continuous path milling was produced. Independent machine tool builders have subsequently developed the systems currently available.

By 1957, the first successful NC installations were being used in production; however, many users were experiencing difficulty in generating part programs for input to machine controller. To remedy this situation, M. I. T. began the development of a computer based part programming language called APT—automatically programmed tools④. The objective was to specify mathematical relationships in a straightforward manner. In 1962 the first APT programming system was released for general industrial applications.

The development of numerical control technology has taken place on two major fronts. Hardware development concentrated on improved control system and machine tools. Sophisticated NC machines and control systems were available by 1965 for every major machine tool configuration. Software development concentrated on improvements to the APT language as well as the origination of other NC programming systems.

A change in overall philosophy began in the 1970s, and numerical control was then viewed as part of a larger concept—computer-aided manufacturing (CAM). CAM encompasses not only NC but production control and monitoring, materials management, and scheduling. The emphasis on the use of computers in the manufacturing process has spawned new forms of numerical control: CNC (computer numerical control) and DNC (direct numerical control). Numerical control continues to develop, and possibilities that were once considered science fiction are now seen as attainable goals.

Numerical control performs best where other forms of specialized automation fail. NC is a system that can interpret a set of prerecorded instructions in some symbolic format; it can cause the controlled machine to execute the instructions, and then can monitor the results so that the required precision and function are maintained.

Numerical control is not a kind of machine tool but a technique for controlling a wide variety of machines. For this reason NC has been applied to assembly machines, inspection equipment, drafting machines, typesetters, woodworking machines (to name only a few applications) as well as metal cutting machine tools. Because NC was originally developed for the machine tool, and because NC metal cutting tools comprise the vast majority of all NC applications, we consider the NC machine tool exclusively in this text.

The numerical control system forms a communication link which has many similarities to conventional processes. A simplified schematic of the most important NC system elements is shown in Fig. 24.1. Symbolic instructions are input to an electronic control unit which decodes them, performs any logical operations required, and outputs precise instructions that control the operation of the machine. Many NC systems contain sensing devices that transmit machine status back to the control unit. It is this feedback that enables the controller to verify that the machine operation conforms to the symbolic input instructions⑤.

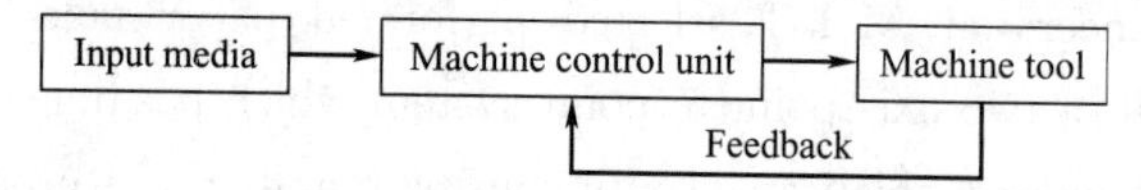

Fig. 24.1 A simplified schematic of an NC system

Selected from "Numerical Control and Computer-Aided Manufacturing", Roger S. Pressman and John E. Williams, John Wiley & Sons, 1977.

New Words and Expressions

1. intervention [ˌintəˈvenʃən] *n.* 介入，干预
2. conceive [kənˈsiːv] *v.* 设想，想象
3. incremental [inkriˈməntəl] *a.* 增加的，增量的，递增的
4. perforate [ˈpəːfəreit] *v.* 冲孔，穿孔
5. position [pəˈziʃən] *v.* 确定……的位置，定位
6. subsequently [ˈsʌbsikwəntli] *ad.* 其次，其后，接着
7. origination [əˌridʒiˈneiʃən] *n.* 产生，出现，发明
8. typesetter [ˈtaipˌsetə] *n.* 字母打印机

Notes

① 参考译文：伴随着新的难于加工的材料的出现以及精确度达万分之一的（加工）误差要求，最优秀的操作者已达到他们能力的极限。

hard-to-machine：难于加工的；one part in ten thousand：万分之一。

② 参考译文：正如他所设想的，坐标点在穿孔卡片上被编码并送到机床控制器中，机床控制器使经改进的铣床以小增量步长运动从而得到所需轨迹。

③ Massachusetts Institute of Technology 麻省理工学院，英文缩写为 M. I. T.。

④ APT—automatically programmed tools 刀具控制程序自动编制系统。

⑤ 参考译文：正是这一反馈使控制器能够验证机床操作与符号式的输入指令相符。

此句为强调句型，强调主语 this feedback。

Exercises

1. After reading the text above, write a summary of it.
2. Answer the following questions according to the text.

① When did people begin to develop the NC machines and what communication medium did they often use?

② Why did scientists and engineers develop APT systems?

③ According to the text, is NC a kind of machine tool or a technique for controlling a wide variety of machines?

④ Why did people use feedback in the NC systems?

3. Translate the last paragraph into Chinese.
4. Put the following into Chinese by reference to the text.

flexibility　literature　straightforward　encompass　spawn　schematic

CNC (computer numerical control)

5. Put the following into English.

缩写　铣　点至点的，点位的　直接数控　解码，译码

6. Translate the following sentences into English.

① 数控程序包括一系列的命令，这些命令使数控机床执行某种操作，机加工是其中最常用的操作工艺。

② 数控系统包括下列三个基本组成部分：指令程序；机床控制装置；工艺设备。

Reading Material 24

Computerized Numerical Control (2)

The Numerically Controlled Machines

We have defined NC as a method of automatic control that uses symbolically coded instructions to cause the machine to perform a specific series of operations. The most important of these instructions is that of position.

Positioning

Although modern NC systems perform many functions, the most important controlled operation is static or dynamic positioning of the cutting tool with the use of a system of coordinates that is general enough to define any geometric motion.

The right-handed Cartesian coordinate provides a simple method for the definition of any point in three-dimensional space. While all NC machines make use of a coordinate system, some require only two axis (x and y) motion, and others require three-dimensional linear and angular axes. For the present purposes the 3-D Cartesian system is used.

Unlike a pure Cartesian system in which an infinite number of axial subdivisions are assumed, a NC coordinate system considers only a finite number of subdivisions. Each command which is output by the control unit to the machine corresponds to one unit of motion. A NC machine is only as accurate as this smallest predefined subdivisions. To illustrate this concept, consider a one-axis NC device which generates motion along the x-axis bounded by the values $x=-0.1$ m and $x=+0.1$ m. If the axis has 1,000 predefined subdivisions, the machine could position to 0.2 mm. If 10,000 subdivisions were used, accuracy would increase tenfold to 0.02 mm.

A NC machine tool uses the coordinate system as a framework for positioning the cutting tool with respect to the workpiece. To accomplish this, points along the component profile (in many cases offset points are required) are defined by x, y, z coordinates. These coordinates are then fed in sequence to the NC controller which generates the appropriate positioning commands. A typical machine-axis configuration is illustrated in Fig. 24.2. Any combination of spindle motion and/or workholding table motion enables the proper position to be attained.

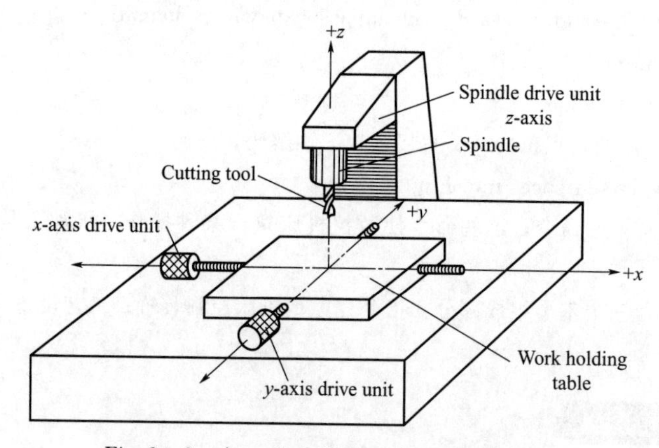

Fig. 24.2 A typical NC machine axis system

Positioning can be accomplished using two distinct methods. The first method, called absolute positioning, fixes the reference system and enables the actual x, y, z coordinates to be specified with reference to a fixed origin. Using an absolute positioning system, the points P_1 and P_2 would be specified as (x_1, y_1, z_1) followed by (x_2, y_2, z_2), regardless of the cutter position before the command is issued.

The second method of positioning uses incremental movement to obtain the same result. In an incremental device, the reference system is relative to the last position.

Control System

In our discussion of positioning, the path taken between points was disregarded. The path which the cutting tool follows as it traverses from point to point depends upon the type of control system used. Three basic path control systems are found in general usage.

The point-to-point system (also called a positioning system) effectively disregards the path between points. Each axis of motion is controlled independently so that the path steps from the start position to the next position as shown in Fig. 24. 3. The path shown is not unique as some point-to-point systems first satisfy the x command and then the y, whereas others reverse the order of execution. Because the traverse path is not controlled, point-to-point systems can only be used in NC applications in which a discrete operation, such as drilling a hole, retracting, drilling another hole, occurs at a given stationary location.

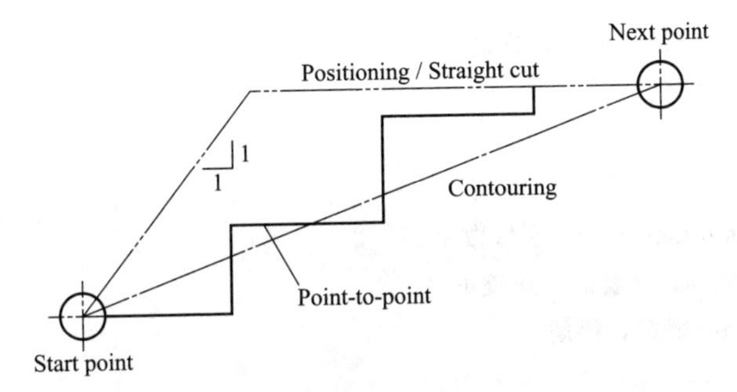

Fig. 24. 3 Comparison of control system paths

Positioning /straight-cut systems provide a somewhat greater degree of axis coordination than point-to-point devices. The straight-cut system has the ability to accurately follow straight paths along each machine axis. Many controllers of this type also produce limited diagonal paths by maintaining a one-to-one relationship among the motions of each axis.

Contouring systems are the most versatile and sophisticated NC devices. The contouring controller generates a path between points by interpolating intermediate coordinates. All contouring systems have a linear interpolation capability (i. e., the ability to generate a straight line between two points). Many systems have additional interpolation capabilities, and because the contouring system provides a predictable, accurate path between points, any path in space can be traced.

Fig. 24. 3 compares the typical motion for each system. The change in position with respect to time must also be considered for successful machining.

Communication Media

A numerically controlled machine will function only if the proper instructions are developed and passed to the machine control. The process by which the symbolic NC instructions are transferred to the control unit is termed the communication cycle. The cycle begins with the development of a set of NC instructions, called a part program, that specifies positioning data and related machining functions in a machine-readable format.

The next step in the communication cycle is the physical transfer of the part program to the machine controller. The communication medium transports a symbolically coded part program to the control unit. Even a relatively short set of NC instructions may contain hundreds, and possibly thousands, of alphanumeric characters and special symbols. For this reason a communication medium must represent symbolic code in a compact form which can be easily deciphered by the machine control.

Selected from "Numerical Control and Computer-Aided Manufacturing", Roger S. Pressman and John E. Williams, John Wiley & Sons, 1977.

New Words and Expressions

1. three-dimensional [θriːdimenʃənl] *a*. 三维的，3D 的
2. subdivision [ˈsʌbdiˌviʒən] *n*. 细分，细分度
3. tenfold [ˈtenfəuld] *a*.；*ad*. 十倍的 [地]
4. profile [ˈprəufail] *n*. 轮廓，断面
5. traverse [ˈtrævəːs] *v*. 横行，横向，横动
6. retract [riˈtrækt] *v*. 缩回，收回
7. diagonal [daiˈægənl] *a*. 对角（线）的，对顶（线）的
8. interpolate [inˈtəːpəuleit] *v*. 插入，内插，插补
9. alphanumeric [ˌælfənjuːˈmerik] *a*. 字母数字的
10. compact [kəmˈpækt] *a*. 紧凑的，压缩的
11. decipher [diˈsaifə] *v*. 解释，译解

Unit 25 ● Robots (1)

Industrial robots became a reality in the early 1960's when Joseph Engelberger and George Devol teamed up to form a robotics company they called "Unimation".

A robot is not simply another automated machine. Automation began during the industrial revolution with machines that performed jobs that formerly had been done by human workers①. Such a machine, however, can do only the specific job for which it was designed, whereas a robot can perform a variety of jobs.

A robot must have an arm. The arm must be able to duplicate the movements of a human worker in loading and unloading other automated machines, spraying paint, welding, and performing hundreds of other jobs that cannot be easily done with conventional automated machines.

Definition of A Robot

The Robot Industries Association (RIA) has published a definition for robots in an attempt to clarify which machines are simply automated machines and which machines are truly robots. The RIA definition is as follows:

A robot is a reprogrammable multifunctional manipulator designed to move material, parts, tools, or specialized devices through variable programmed motions for the performance of a variety of tasks②.

This definition, which is more extensive than the one in the RIA glossary at the end of this book, is an excellent definition of a robot. We will look at this definition, one phrase at a time, so as to understand which machines are in fact robots and which machines are little more than specialized automation.

First, a robot is a "reprogrammable multifunctional manipulator." In this phrase RIA tells us that a robot can be taught (reprogrammed) to do more than one job by changing the information stored in its memory. A robot can be reprogrammed to load and unload machines, weld, and do many other jobs (multifunctional). A robot is a "manipulator". A manipulator is an arm (or hand) that can pick up or move things. At this point we know that a robot is an arm that can be taught to do different jobs.

The definition goes on to say that a robot is "designed to move material, parts, tools, or specialized devices". Material includes wood, steel, plastic, cardboard... anything that is used in the manufacture of a product.

A robot can also handle parts that have been manufactured. For example, a robot can load a piece of steel into an automatic lathe and unload a finished part out of the lathe.

In addition to handling material and parts, a robot can be fitted with tools such as grinders, buffers, screwdrivers, and welding torches to perform useful work.

Robots can also be fitted with specialized instruments or devices to do special jobs in a manufacturing plant. Robots can be fitted with television cameras for inspection of parts or products. They can be fitted with lasers to accurately measure the size of parts being manufactured.

The RIA definition closes with the phrase, "... through variable programmed motions for the performance of a variety of tasks." This phrase emphasizes the fact that a robot can do many different jobs in a manufacturing plant. The variety of jobs that a robot can do is limited only by the creativity of the application engineer.

Jobs for Robots

Jobs performed by robots can be divided into two major categories: hazardous jobs and repetitive jobs.

Hazardous Jobs

Many applications of robots are in jobs that are hazardous to humans. Such jobs may be considered hazardous because of toxic fumes, the weight of the material being handled, the temperature of the material being handled, the danger of working near rotating or press machinery, or environments containing high levels of radiation.

Repetitive Jobs

In addition to taking over hazardous jobs, robots are well suited to doing extremely repetitive jobs that must be done in manufacturing plants③. Many jobs in manufacturing plants require a person to act more like a machine than like a human. The job may be to pick a piece up from here and place it there. The same job is done hundreds of times each day. The job requires little or no judgment and little or no skill. This is not said as a criticism of the person who does the job, but is intended simply to point out that many of these jobs exist in industry and must be done to complete the manufacture of products. A robot can be placed at such a work station and can perform the job admirably without complaining or experiencing the fatigue and boredom normally associated with such a job.

Selected from "Robotics and Automated Systems", Robert L. Hoekstra, CmfgE, South-Western Publishing Co., 1986.

New Words and Expressions

1. unimation [ju:ni'meiʃən] *n*. 通用机械手（一种机器人的商品名）
2. duplicate ['dju:plikeit] *v*. 复制
3. multifunctional [ˌmʌlti'fʌŋkʃənl] *a*. 多功能的
4. manipulator [mə'nipjuleitə] *n*. 机械手，操作手，操纵器
5. glossary ['glɔsəri] *n*. 术语集，词集，专业词典
6. grinder ['graində] *n*. 磨床
7. buffer ['bʌfə] *n*. 缓冲器，减震器，保险杠，抛光轮
8. screwdriver ['skru:draivə] *n*. 螺丝起子，改锥，旋凿，螺丝刀
9. hazardous ['hæzədəs] *a*. 危险的，有害的

Notes

① 参考译文：自动化始于工业革命时期，它用机器去完成早先由人工去做的作业。

② 参考译文：机器人是一种可重新编程的、多功能的机械手，为实现各种任务设计成通过可改变的程序动作来移动材料、零部件、工具或专用装置。

③ 参考译文：除了担任危险工作外，机器人也非常适合做那些在制造厂里必须做的完全重复的工作。

Exercises

1. After reading the text above，summarize the main ideas of it in oral English.
2. Answer the following questions according to the text.
   ① Please list differences between automated machine and robot.
   ② What is a robot?
   ③ What are hazardous jobs for robots?
   ④ What are repetitive jobs for robots?
3. Translate the 7th and 15th paragraphs of the text into Chinese.
4. Put the following into Chinese by reference to the text.
   manipulator　lathe　multifunctional　hazardous　repetitive
5. Put the following into English.
   疲劳　机器人　磨床　焊接　机械手
6. Translate the following sentences into English.
   ① 机器人执行的工作可分为两大类：危险性工作和重复性工作。
   ② 机器人是一种可重新编程的多功能的机械手，用来移动材料、零部件和工具。

Reading Material 25

Robots (2)

Robot Speed

Although robots increase productivity in a manufacturing plant，they are not exceptionally fast. At present，robots normally operate at or near the speed of a human operator. Every major move of a robot normally takes approximately one second. For a robot to pick up a piece of steel from a conveyor and load it into a lathe may require ten different moves taking as much as ten seconds. A human operator can do the same job in the same amount of time. The increase in productivity is a result of the consistency of operation. As the human operator repeats the same job over and over during the workday，he or she begins to slow down. The robot continues to operate at its programmed speed and therefore completes more parts during the workday.

Custom-built automated machines can be built to do the same jobs that robots do. An automated machine can do the same loading operation in less than half the time required by a robot or a human operator. The problem with designing a special machine is that such a machine can perform only the specific job for which it was built. If any change is made in the job, the machine must be completely rebuilt, or the machine must be scrapped and a new machine designed and built. A robot, on the other hand, could be reprogrammed and could start doing the new job the same day.

Custom-built automated machines still have their place in industry. If a company knows that a job will not change for many years, the faster custom-built machine is still a good choice.

Other jobs in factories cannot be done easily with custom-built machinery. For these applications a robot may be a good choice. An example of such an application is spray painting. Spray painting is a hazardous job, because the fumes from many paints are both toxic and explosive. A robot is now doing the job of spraying paint on the enclosures. A robot has been "taught" to spray all the different sizes of enclosures that the company builds. In addition, the robot can operate in the toxic environment of the spray booth without any concern for the long-term effect the fumes might have on a person working in the booth.

Flexible Automation

Robots have another advantage: they can be taught to do different jobs in the manufacturing plant. If a robot was originally purchased to load and unload a punch press and the job is no longer needed due to a change in product design, the robot can be moved to another job in the plant. For example, the robot could be moved to the end of the assembly operation and be used to unload the finished enclosures from a conveyor and load them onto a pallet for shipment.

Accuracy and Repeatability

One very important characteristic of any robot is the accuracy with which it can perform its task. When the robot is programmed to perform a specific task, it is led to specific points and programmed to remember the locations of those points. After programming has been completed, the robot is switched to "run" and the program is executed. Unfortunately, the robot will not go to the exact location of any programmed point. For example, the robot may miss the exact point by 0.025 in. If 0.025 in. is the greatest error by which the robot misses any point during the first execution of the program, the robot is said to have an accuracy of 0.025 in.

In addition to accuracy, we are also concerned with the robot's repeatability. The repeatability of a robot is a measure of how closely it returns to its programmed points every time the program is executed. Say, for example, that the robot misses a programmed point by 0.025 in. the first time the program is executed and that, during the next execution of the program, the robot misses the point it reached during the previous cycle by 0.010 in. Although the robot is a total of 0.035 in. from the original programmed point, its accuracy is 0.025 in. and its repeatability is 0.010 in.

The Major Parts of a Robot

The major parts of a robot are the manipulator, the power supply, and the controller.

The manipulator is used to pick up material, parts, or special tools used in manufacturing. The power supply supplies the power to move the manipulator. The controller controls the power supply so that the manipulator can be taught to perform its task.

Axes of Robot Movement

The various movements that the manipulator of a robot can make are defined by its degrees of freedom or axes. If a robot's manipulator can rotate, the robot is said to be a single-axis robot. If the manipulator can move up and down as well as rotate, the robot is called a two-axis robot. If, in addition to the rotational movement and the up-and-down movement, the manipulator can also extend its arm, or "reach", the robot is said to be a three-axis robot (Fig. 25. 1). Most industrial robots have all three major axes (rotational, up and down, and reach) as well as some minor axes of movement.

The minor axes of a robot are found in the robot's wrist. The wrist of a robot is attached to the end of the robot's arm. There are three possible movements or axes of a robot wrist: pitch, roll, and yaw (Fig. 25. 2). The pitch movement bends the wrist up and down. The roll movement is the twisting of the wrist. The yaw movement is the side-to-side movement of the wrist.

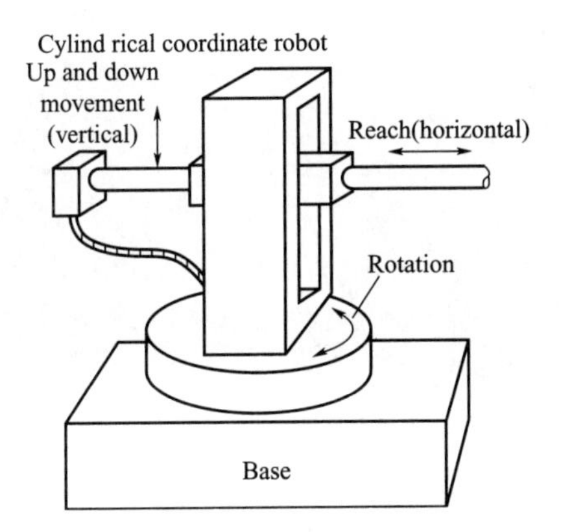

Fig. 25. 1 Three-axis robot with cylindrical (post-type) manipulator, illustrating two linear axes, and one rotational axis, of movement

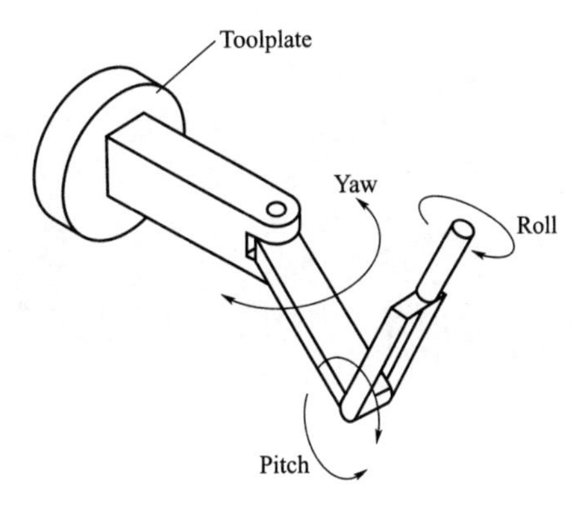

Fig. 25. 2 The three possible movements or axes of a robot wrist

The combination of the major axes and the minor axes gives the robot six possible movements (six axes or six degrees of freedom). Many industrial robots are equipped with all six axes. Some robots, however, have the three major axes but only one or two of the minor axes.

Classification of Robots

The total area that the end of the robot's arm can reach is called the work envelope. Robots can be classified according to their work envelopes into four types: the cylindrical coordinate robot, the rectangular coordinate robot, the spherical coordinate robot, and the jointed arm coordinate robot. They also can be classified by motion control. There are three major classifications of motion for robots: pick-and-place, point-to-point, and continuous path. These terms describe the movements the manipulator can make within its work envelope.

Selected from "Robotics and Automated Systems", Robert L. Hoekstra, CmfgE, South-Western Publishing Co., 1986.

New Words and Expressions

1. approximately [əprɔksi'mətli] *ad*. 大约，大致，近于
2. consistency [kən'sistənsi] *n*. 一致性
3. scrap [skræp] *v*. 扔弃，敲碎，拆毁，报废
4. fume [fjuːm] *n*. 烟雾，烟
5. toxic ['tɔksik] *a*. 有毒的，中毒的
6. booth [buːθ] *n*. 亭，摊棚，摊位
7. execution [ˌeksi'kjuːʃən] *n*. 执行，实行
8. pitch [pitʃ] *n*. 俯仰，倾斜，齿距，螺距，节距
9. be equipped with 装备着，安置着
10. work envelope 工作范围
11. cylindrical [si'lindrikəl] *a*. 圆柱形的，圆筒形的，柱面的
12. spherical ['sferikəl] *a*. 球形的，球状的

Unit 26 ● Computer-Aided Manufacturing (1)

A computer-aided manufacturing (CAM) system oversees many aspects of manufacture by introducing a hierarchical computer structure to monitor and control various phases of the manufacturing process. Conventional and adaptive NC systems are the predecessors to larger CAM systems. Where NC considers an information feedback loop concerned with a discrete process, CAM develops an integrated information network that monitors a broad spectrum of interrelated tasks and controls each based on an overall management strategy①.

Ideally, a CAM system should have three attributes applicable to each phase of the manufacturing process: although supervised, a minimum amount of human intervention should be required for individual process tasks; the system should be flexible and allow processes to be individually programmed; and the CAM system should be integrated, with both engineering design and analysis, using a computer-aided design (CAD) system. Various aspects of this ideal system have been successfully implemented, but much development work continues to be carried out.

The CAM Hierarchy

A large scale CAM system contains a hierarchical structure of two or three levels of computers that are used to control and monitor individual process tasks. A small (mini-) computer is responsible for the management of a single processed information. This general configuration of a CAM hierarchy is illustrated in Fig. 26. 1.

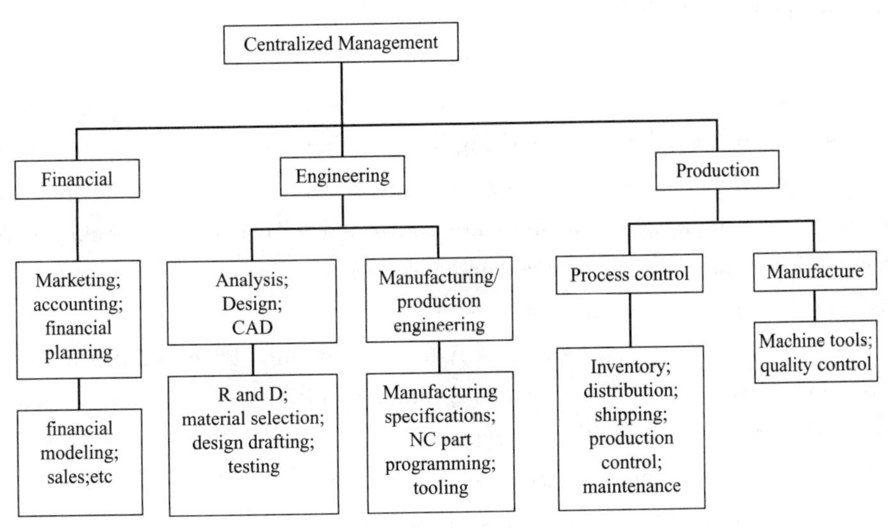

Fig. 26. 1 Computer-aided manufacturing (CAM) hierarchical structure

A large scale CAM system encompasses three major areas related to the manufacturing process: production management and control; engineering analysis and design; and finance and marketing②. Each is comprised of subtasks that are controlled either directly from a

large computer (e. g. , inventory control), or by a small computer, as in the case of inspection/quality control. Regardless of the control method, the important strength of CAM is that a two-way flow of information occurs.

Because the CAM system oversees many aspects of the manufacturing process, changes dictated by information monitored from one subtask can be translated into control data for some other subtask③. For example, in the manufacturing task, machining, inspection, and assembly are all under computer control. When the computer recognizes that a component is continually out of tolerance (based on information feedback from automated testing equipment), it can be programmed to effect a change in the actual machining process to compensate for the error. Since both subtasks become part of the same information loop, a feedback control system including machining and inspection is established.

NC in Computer-Aided Manufacturing

Although numerically controlled machine tools are essential for the development of operational CAM systems, those described in this text cannot be used in a computer based manufacturing system. Conventional NC and AC machines must be modified so that information may be passed between the MCU and a computer based system. This modification has result in three major developments derived from the NC concept: computer managed numerical control, the cluster concept, and new forms of adaptive control.

Computer managed numerical control is a generic term that encompasses DNC (Direct Numerical Control) and CNC (Computerized Numerical Control). DNC and CNC are methods that distribute programmable computing responsibility between a control computer and NC machine tool. Neither system changes the functional characteristics of the NC machine; instead, each provides a means for communication of process data and commands outside the NC machine control loop④.

The cluster concept is essentially an extension of computer management to more than one kind of machine⑤. A series of machine tools (e. g. , those used for milling, boring, grinding) are interconnected by a conveying system that automatically supplies individual machines with components at the required time. Two levels of control and monitoring become necessary. The individual machines are controlled with computer managed NC, and the cluster itself is managed by a centralized computer coordinating the production output of many clusters. Again, a hierarchical arrangement becomes evident.

Adaptive NC systems are part of the CAM environment. Process information is made available to centralized computer so that exceptional conditions (e. g. , tool breakage) may be detected and corrected. In addition, adaptive feedback may also be recorded and analyzed so that the production efficiency of a given operation may be established.

The use of NC machines in a computer based manufacturing system can be viewed in terms of automation and integration. Referring to Fig. 26. 2, four levels of manufacturing automation can be defined. The stand alone NC machine tool represents an automated operating cycle; an NC machining center automates the entire machining process; a cluster or group of externally controlled NC machines represents a fully automated manufacturing

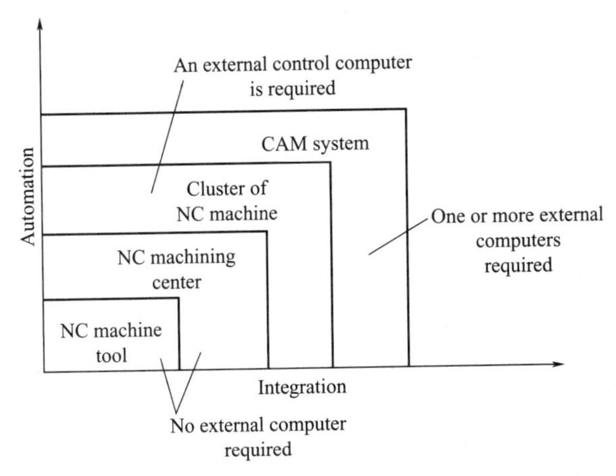

Fig. 26. 2 Integration and automation for numerical control in CAM

task, and finally, the CAM system itself integrates all lower level methods in an automated manufacturing process⑥.

Selected from "Numerical Control and Computer-Aided Manufacturing", Roger S. Pressman and John E. Williams, John Wiley & Sons, 1977.

New Words and Expressions

1. oversee [ˈəuvəˈsiː] *v*. **管理，照料**
2. predecessor [ˈpriːdisesə] *n*. **（被代替的）原有（事）物，前任**
3. spectrum [ˈspektrəm] *a*. **范围，系列，领域，谱（如：光谱、频谱等）**
4. hierarchy [ˈhaiərɑːki] *n*. **分级结构，层次，等级制度**
5. R and D = research and development **研究与发展，研制与试验，研究与开发（研发）**
6. inventory [ˈinvəntri] *n*. **库存，存货（清单），报表**
7. dictate [dikˈteit] *v*. **命令，支配**
8. MCU=monitor and control unit **监控设备，监控装置**
9. cluster [ˈklʌstə] *n*. **群，组**

Notes

① 参考译文：数控（NC）在考虑与离散过程有关的信息反馈回路处，计算机辅助制造（CAM）开发了一种集成信息网络，此网络可监视范围广阔相互关联的各种任务，并按总体管理策略控制每项任务。

② 参考译文：一个大规模的CAM系统包含三个与制造工艺过程有关的主要领域：生产管理与控制，工程分析与设计，财务与市场。

③ 参考译文：因为CAM系统管理制造工艺过程的很多方面，由一个子任务监测到的信息所支配的变化，能够转换为其他子任务的控制数据。

④ 参考译文：两个系统都不改变数控机床的功能特性，而是每一系统都提供了在数控机床控制回路之外的工艺过程数据与命令的通讯方法。

⑤ 参考译文：群集概念实质上是对多种机床的计算机管理的延伸。

⑥ 参考译文：单独使用的数控机床表示某一自动化操作过程；数控机加工中心使整个

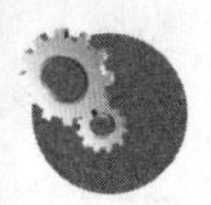

机加工工艺过程（实现）自动化；一组外部（计算机）控制的数控机床表示某一完全自动化的制造任务，而最后，CAM 本身集成了所有在自动化制造过程工艺中的较低级水平的方法。

Exercises

1. After reading the text above，write a summary of it.
2. Answer the following questions according to the text.
   ① What are the three attributes a CAM system should have?
   ② Describe the three major areas related to the manufacturing process a large scale CAM system has.
   ③ What does "those"（in line 2，paragraph 6）refer to?
   ④ Give an example of the application of adaptive NC system.
3. Translate the 5th paragraph into Chinese.
4. Put the following into Chinese by reference to the text.
   discrete　implement　configuration　specification　regardless of　AC
   in terms of
5. Put the following into English.
   计算机辅助制造　适用于　子任务　双向的
   超出公差（范围）　补偿　向 M 提供 N
6. Translate the following sentences into English.
   ① CAD/CAM 是指在设计和生产过程中使用数字计算机完成某些操作的技术。
   ② CAM 的直接应用是指将计算机直接与制造过程连接以对它进行监视和控制。

Reading Material 26

Computer-Aided Manufacturing (2)

Elements of the CAM System

The success of computer-aided manufacturing systems depends upon integration of hardware and software functioning in the overall information flow. CAM hardware elements include NC machine tools，inspection equipment，digital computers，and related devices. CAM software is an interrelated mesh of computer programming systems that serve to monitor，process，and ultimately control the flow of manufacturing data and CAM hardware. In this section the software elements of the CAM system are examined in relation to the flow of information.

The CAM Data Base

A computer based manufacturing system relies upon both real time and stored information that is readily available in an accessible form. Real time data，such as AC feedback，is generally processed immediately and，except for record keeping purposes，can be considered transient information. Stored data provides the CAM system with all necessary input required to perform control and analysis functions. All forms of stored data are maintained in a data base that can be accessed at extremely high speed by computer.

The level of complexity of a CAM data base is directly proportional to the number of tasks required by the system. In the ideal CAM system (Fig. 26. 1) an extremely large and complex data base configuration is required. For an attainable CAM system (by present standards), the data base should contain those elements illustrated in Fig. 26. 3.

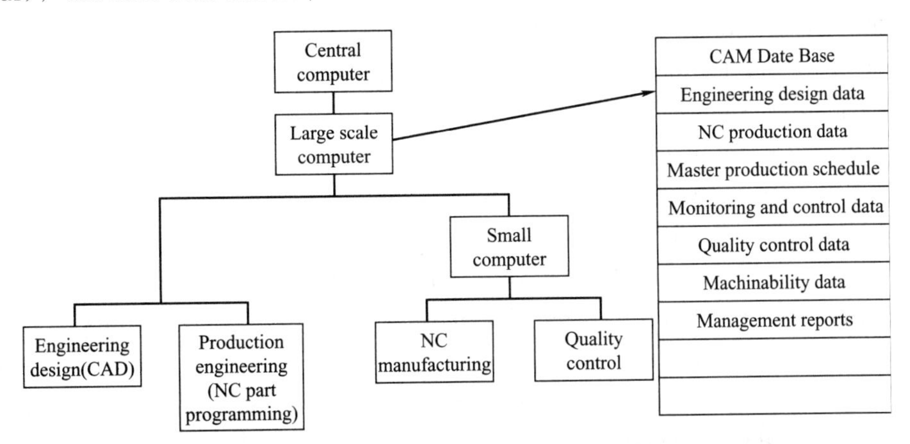

Fig. 26. 3 CAM data base for a computer based manufacturing system

For the CAM system shown, only the engineering and manufacturing functions are under direct system control. Engineering design data enters the data base from computer-aided design programs used by various engineering departments. The design data and externally supplied manufacturing specifications are used by production/manufacturing engineering to develop NC part programs and other operational specifications. These are stored as NC production information in the data base.

Production Management

Once the finalized NC part program has been stored in the CAM data base, manufacturing priority dictates when the program is executed. The master production schedule, developed to reflect priority requirements of an individual company, is used by a scheduling program to control production. The scheduling program receives information concerning the status of production and makes the following information available:

① Status of individual parts undergoing machining operations;

② Status of each NC machine tool;

③ Actual production times versus scheduled production times;

④ Impending machine or system failures.

Based on this information, the scheduling program determines the production load for each operational machine tool so that established priorities are maintained.

When the scheduling program determines which NC part program is to be executed next, it places the data in a ready bank that is accessible, either directly or indirectly, to the numerical control machine's MCU. The program may also generate information concerning tooling, setup times, and manufacturing time for use by production personnel.

Manufacturing Control

The manufacturing control mode depends on the type of NC configuration used in the

CAM system. Consider the generalized manufacturing control system. The manufacturing control program can be executed by a minicomputer at the NC machine site (CNC), a real time computer controlling a group of NC machines, or a large scale computer being linked to the NC machine tool via telecommunications lines (DNC).

The manufacturing control program passes instructions to the NC machine based on data currently in the ready bank. NC data is obtained in blocks (subdivisions of the entire data structure or part program) from this ready bank as required. Individual NC blocks of information are passed to the machine. Process information and machine status are conveyed by the control program to the appropriate data evaluation modules.

The quality control function uses component specifications obtained from the CAM data base. The quality control program may receive manual input via interactive terminals at the inspection site or data obtained from NC inspection equipment. Regardless of the data source, the quality control program performs two distinct functions: it provides accurate validation of manufactured parts and initiates action by the manufacturing control system when tolerance exceptions occur.

Selected from "Numerical Control and Computer-Aided Manufacturing", Roger S. Pressman and John E. Williams, John Wiley & Sons, 1977.

New Words and Expressions

1. transient ['trænziənt] *a*. 瞬时的，瞬变的
2. master production schedule 总生产进度表
3. impending [im'pendiŋ] *a*. 迫切的，即将发生的
4. ready bank 现成数据库就绪数据库
5. evaluation [iˌvælju'eiʃən] *n*. 已取得数据的整理，求值，赋值，评估，估（测）算

Unit 27 • Flexible Manufacturing Systems (1)

Flexible Manufacturing System Defined

The evolution of manufacturing can be represented graphically as a continuum as shown in Fig. 27. 1. As this figure shows, manufacturing processes and systems are in a state of transition from manual operation to the eventual realization of fully integrated manufacturing. The step preceding computer-integrated manufacturing is called flexible manufacturing.

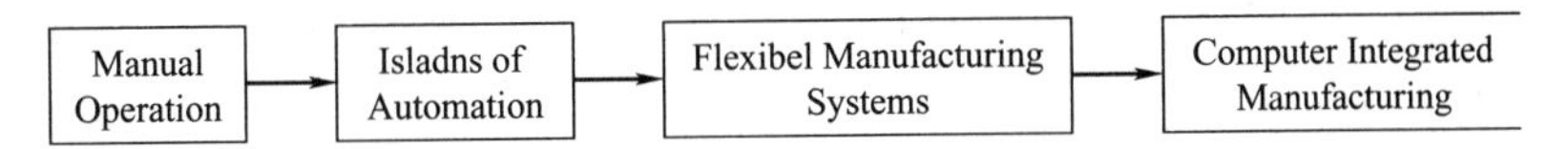

Fig. 27. 1 The evolution of manufacturing

Flexibility is an important characteristic in the modern manufacturing setting. It means that a manufacturing system is versatile and adaptable, while also capable of handling relatively high production runs. A flexible manufacturing system is versatile in that it can produce a variety of parts. It is adaptable in that it can be quickly modified to produce a completely different line of parts. This flexibility can be the difference between success and failure in a competitive international marketplace.

It is a matter of balance. Stand-alone computer numerical control (CNC) machines have a high degree of flexibility, but are capable of relatively low-volume production runs. At the opposite end of the spectrum, transfer lines are capable of high-volume runs, but they are not very flexible. Flexible manufacturing is an attempt to use technology in such a way as to achieve the optimum balance between flexibility and production runs①. These technologies include automated materials handling, group technology, and computer and distributed numerical control.

A flexible manufacturing system (FMS) is an individual machine or group of machines served by an automated materials handling system that is computer controlled and has a tool handling capability②. Because of its tool handling capability and computer control, such a system can be continually reconfigured to manufacture a wide variety of parts③. This is why it is called a flexible manufacturing system.

The key elements necessary for a manufacturing system to qualify as an FMS are as follows:

① computer control;

② automated materials handling capability;

③ tool handling capability.

Flexible manufacturing represents a major step toward the goal of fully integrated manufacturing in that it involves integration of automated production processes. In flexible manufacturing, the automated manufacturing machine (i. e. , lathe, mill, drill) and the auto-

mated materials handling system share instantaneous communication via a computer network. This is integration on a small scale.

Overview of Flexible Manufacturing

Flexible manufacturing takes a major step toward the goal of fully integrated manufacturing by integrating several automated manufacturing concepts:

① Computer numerical control (CNC) of individual machine tools;

② Distributed numerical control (DNC) of manufacturing systems;

③ Automated materials handling systems;

④ Group technology (families of parts).

When these automated processes, machines, and concepts are brought together in one integrated system, an FMS is the result. Humans and computers play major roles in an FMS. The amount of human labor is much less than with a manually operated manufacturing system, of course. However, humans still play a vital role in the operation of an FMS. Human tasks include the followings:

① Equipment troubleshooting, maintenance, and repair;

② Tool changing and setup;

③ Loading and unloading the system;

④ Data input;

⑤ Changing of parts programs;

⑥ Development of programs.

Flexible manufacturing system equipment, like all manufacturing equipment, must be monitored for "bugs", malfunctions, and breakdowns. When a problem is discovered, a human troubleshooter must identify its source and prescribe corrective measures. Humans also undertake the prescribed measures or repair the malfunctioning equipment. Even when all systems are properly functioning, periodic maintenance is necessary.

Human operators also set up machines, change tools, and reconfigure systems as necessary. The tool handling capability of an FMS decreases, but does not eliminate, human involvement in tool changing and setup. The same is true of loading and unloading the FMS. Once raw material has been loaded onto the automated materials handling system, it is moved through the system in the prescribed manner. However, the original loading onto the materials handling system is still usually done by human operators, as is the unloading of finished products.

Humans are also needed for interaction with the computer. Humans develop parts programs that control the FMS via computers. They also change the programs as necessary when reconfiguring the FMS to produce another type of part or parts. Humans also input data needed by the FMS during manufacturing operations. Humans play less labor-intensive roles in an FMS, but the roles are still critical.

Control at all levels in an FMS is provided by computers. Individual machine tools within an FMS are controlled by CNC. The overall system is controlled by DNC. The automated materials handling system is computer controlled, as are other function including data colle-

ction, system monitoring, tool control, and traffic control. Human/computer interaction is the key to the flexibility of an FMS.

Selected from "Advanced Manufacturing Technology", David L. Goetsch, Delmar Publishers Inc., 1990.

New Words and Expressions

1. flexible manufacturing 柔性制造
2. continuum [kən'tinjuəm] *n*. 连续（统一体）
3. transition [træn'ziʃən] *n*. 转变，转换，变迁，过渡时期
4. realization [ˌriəlai'zeiʃən] *n*. 实现
5. stand-alone *a*. 可独立应用的，单独的
6. reconfigure [ˌriːkən'figə] *v*. 重新配置［组合］
7. volume production 批量［成批］生产，量产
8. transfer line 组合机床自动线
9. family of parts 相似部件，零件族［组］
10. bug [bʌg] *n*. 缺陷，困难，错误
11. periodic [piəri'ɔdik] *a*. 周期的，定期的，循环的
12. troubleshooting ['trʌblʃuːtiŋ] *n*. 发现并修理故障，排除故障
13. troubleshooter ['trʌblʃuːtə] *n*. 故障检修工
14. labor-intensive 劳动强度大的，劳动密集的

Notes

① 参考译文：柔性制造试图按这样一种方式来使用技术，以取得灵活性与流水生产之间的最佳平衡。

② 参考译文：柔性制造系统是由计算机控制的自动物料搬运系统服务的单台机床或一组机床，并拥有刀具装卸能力。

③ 参考译文：由于它的刀具装卸能力和计算机控制，所以能够不断重新配置这样的系统去制造种类繁多的零件。

a (considerable, great, wide) variety of 各种各样的，种种的，种类［品种］繁多的。

Exercises

1. After reading the text above, write a summary of it.
2. Answer the following questions according to the text.
   ① What is a flexible manufacturing system (FMS)?
   ② What are the key elements that make a manufacturing system an FMS?
   ③ What four concepts are integrated in an FMS?
   ④ List and explain at least four tasks humans accomplished in an FMS.
3. Translate the 2nd paragraph into Chinese.
4. Put the following into Chinese by reference to the text.
   automated materials handling system　transfer line　unload　family of parts
   prescribed manner　play a vital role
5. Put the following into English.

柔性制造系统　　分布式数字控制　　批量生产　　流水线生产
人工操作的机器系统

6. Translate the following sentences into English.

① 在柔性制造系统中，虽然人类的劳动强度降低了，但人类仍发挥着关键作用。

② 制造工艺和系统正在从人工操作向最终实现完全集成制造转变。

Reading Material 27

Flexible Manufacturing Systems (2)

Rationale for Flexible Manufacturing

In manufacturing there have always been trade-offs between production rates and flexibility. At one end of the spectrum are transfer lines capable of high production rates, but low flexibility. At the other end of the spectrum are independent CNC machines that offer maximum flexibility, but are only capable of low production rates (Fig. 27.2). Note from the figure that flexible manufacturing falls in the middle of the continuum. There has always been a need in manufacturing for a system that could produce higher volume and production runs than could independent machines while still maintaining flexibility.

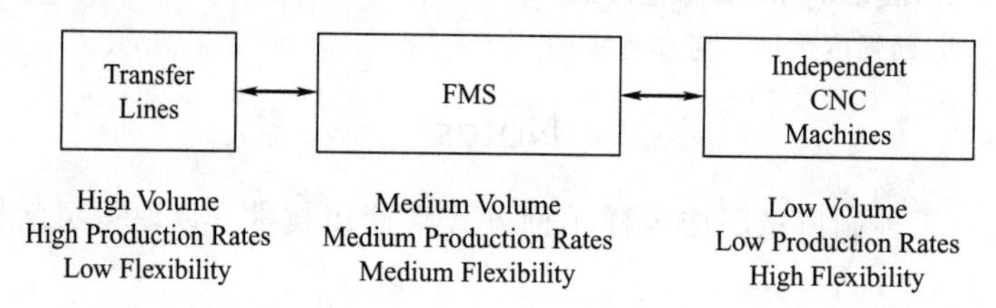

Fig. 27.2　System manufacturing continuum

Transfer lines are capable of producing large volumes of parts at high production rates. The line takes a great deal of setup, but can turn out identical parts in large quantities. Its chief shortcoming is that even minor design changes in a part can cause the entire line to be shut down and reconfigured. This is a critical weakness because it means that transfer lines cannot produce different parts, even parts from within the same family, without costly and time-consuming shutdown and reconfiguration.

Traditionally, CNC machines have been used to produce small volumes of parts that differ slightly in design. Such machines are ideal for this purpose because they can be quickly reprogrammed to accommodate minor or even major design changes. However, as independent machines, they cannot produce parts in large volumes or at high production rates.

An FMS can handle higher volumes and production rates than independent CNC machines. They cannot quite match such machines for flexibility, but they come close. What is particularly significant about the middle ground capabilities of flexible manufacturing is that most manufacturing situations require medium production rates to produce medium volumes with enough flexibility to quickly reconfigure to produce another part or product. Flexible manufacturing fills this longstanding void in manufacturing.

Flexible manufacturing, with its middle ground capabilities, offers a number of advantages for manufacturers:

① Flexibility within a family of parts;

② Random feeding of parts;

③ Simultaneous production of different parts;

④ Decreased setup times/lead time;

⑤ Efficient machine usage;

⑥ Decreased direct and indirect labor costs;

⑦ Ability to handle different materials;

⑧ Ability to continue some production if one machine breaks down.

Flexible Manufacturing System Components

An FMS has four major components:

① Machine tools;

② Control system;

③ Materials handling system;

④ Human operators.

Machine Tools

A flexible manufacturing system uses the same types of machine tools as any other manufacturing system, be it automated or manually operated. These include lathes, mills, saws, and so on. The type of machine tools actually included in an FMS depends on the setting in which the system will be used. Some FMS are designed to meet a specific, well-defined need. In these cases, the machine tools included in the system will be only those necessary for the planned operations. Such a system would be known as a dedicated system.

In a job shop setting, or any other setting in which the actual application is not known ahead of time or must necessarily include a wide range of possibilities, machines capable of performing at least the standard manufacturing operations would be included. Such systems are known as general-purpose systems.

Control Systems

The control system for an FMS serves a number of different control functions for the system:

① storage and distribution of part programs;

② work flow control and monitoring;

③ production control;

④ system/tool control/monitoring.

Materials Handling System

The automated materials handling system is a fundamental component that helps mold a group of independent CNC machines into a comprehensive FMS. The system must be capable of accepting workpieces mounted on pallets and moving them from workstation to workstation as needed. It must also be able to place workpieces "on hold" as they wait to be processed at a given workstation.

The materials handling system must be able to unload a workpiece at one station and load another for transport to the next station. It must accommodate computer control and be completely compatible in that regard with other components in the flexible manufacturing system. Finally, the materials handling system for an FMS must be able to withstand the rigors of a shop environment.

Some FMSs are configured with automated guided vehicles (AGVs) as a principal means of materials handling.

Human Operators

The final component in an FMS is the human component. Although flexible manufacturing as a concept decreases the amount of human involvement in manufacturing, it does not eliminate it completely. Further, the roles humans play in flexible manufacturing are critical roles.

Selected from "Advanced Manufacturing Technology", David L. Goetsch, Delmar Publishers Inc., 1990.

New Words and Expressions

1. trade-off *n*. 折中（办法，方案），权衡，综合
2. time-consuming [ˈtaimkənˈsjuːmiŋ] *a*. 费时（间）的，拖延时间的
3. traditional [trəˈdiʃənl] *a*. 传统的，惯例的
4. accommodate [əˈkɔmədeit] *v*. 使适应
5. longstanding *a*. 长期间的，长期存在的
6. simultaneous [ˈsiməlˈteinjəs] *a*. 同时的，同时发生的
7. lead time 产品设计至实际投产间的时间，提前期，生产准备期
8. dedicated [ˈdedikeitid] *a*. 专用的
9. workpiece *n*. 工件
10. rigor [ˈrigə] *n*. 艰苦，酷烈，严格
11. automated guided vehicles (AGV) 自动导向小车
12. be compatible with M 与M相容（相适应，一致，相似），适合于M

Unit 28 • Computer-Integrated Manufacturing (1)

International Competition and Manufacturing Industry

International competition has intensified the requirement for high quality products that can compete in the global marketplace. As a result of this increased competition, the pace of product or system development has been quickened, thus forcing manufacturer into an era in which continuous quality improvement is a matter of survival, not simply competitive advantage①. As the time scale of the product life cycle has decreased and the demand for quality increased, attention has focused on improving product quality and promoting the competitive ability of companies through better design, manufacturing, management and marketing. Obviously, increasing manufacturing rate and improving product quality will become the most important factors for sharing the international marketplace.

Global manufacturing industry now is undergoing a rapid structural change. As this process continuous, our manufacturing industry is encountering difficulties as it confronts a changed and more competitive environment and marketplace. Clearly, we cannot maintain our industrial base and standard of living without an efficient manufacturing industry.

To join the global competition, to share the international marketplace, as well as to increase our total factor productivity (TFP) and manufacturing labor productivity (MLP), our manufactures may implement two strategies. The first is to change the way we do things now, i. e. to improve our enterprise management. The second one is to develop and apply high technology, i. e. computer-integrated manufacturing (CIM) technology and integrated information management (IIM) for our companies and enterprises. Computer-aided design (CAD) defines and describes products on a video screen; computer-aided engineering (CAE) analyzes production performance and productivity, and computer-aided manufacturing (CAM) automates the shop floor process. As a result, faster, cheaper, safer and better production and operation can be achieved.

Development of Manufacturing Industry

It is widely rccognized that the course of industrialization has substantially been the process of automation. From a viewpoint of automation, the development of industry automation can be divided into four stages as shown in Table 28. 1.

Table 28. 1 Development of industry automation

Stage	Feature	Automation	Design	Manufacture
1	Labor-intensive	None	Individual	Manual
2	Equipment-intensive	Instruments	Group	NC, CNC
3	Information-intensive	Information	CAD	FMS
4	Knowledge-intensive	Decision	ICAD	CIMS

Stage 1. Labor-intensive industry: At this stage, the efficiency and quality of production mainly rely on the skills of human operators using simple machines without automatic controls and operations. Equipment maintenance heavily depends on 'private' experience.

Stage 2. Equipment-intensive industry: Automatic equipment plays a dominant role in the competition of productivity. The equipment may consist of quite competition mechanical, electronic and computerized devices, and can be automatically used only as a stand-alone machine.

A typical example is the numerical control (NC) machine that is highly automatic as a single machine. Meanwhile, sensor techniques and advanced instruments have become important means to obtain industrial data for equipment maintenance and automatic control. As a result of more powerful and affordable computing facilities on factory floors, our industry is now moving into the third stage.

Stage 3. Information-intensive industry: This stimulates the development of flexible manufacturing systems (FMS). At this stage automation is realized at the level of data processing for groups of automatic machines. CAD/CAM technology is a typical example representing the characteristics of the stage. Based on mathematical models, computers assist human experts for numerical analysis, synthesis, simulation and graphics, and also provide information for them to make decisions②. Fault diagnosis and equipment maintenance depend not only on the information from sensors but also on human experts experience.

With the rapid growth of complexity of manufacturing processes and the demand for higher efficiency, greater flexibility, better product quality, and lower cost, industrial practice has approached the more advanced level of automation. Nowadays, much attention has been given by both industry and academia to computer-integrated manufacturing systems (CIMS). What CIMS try to do is to integrate all stages in a product life cycle, including taking orders from customers, production planning, product design, process scheduling, NC coding, product testing, sales, service and production management, into a complete (or integrated) information processing system with minimum interference to the operation process by human operators③.

Stage 4. Knowledge-intensive industry: At this stage, computers help human experts not only for data processing, but also for decision-making. The tendency to replace human brain power by computers is developing into decision-making automation, which is based on the technology of knowledge process. In future industrial companies, intelligent computer-aided design (ICAD) will automatically generate and analyze product models with the help of human expertise; CIMS technology will be used to implement plant-wide integration; and intelligent maintenance systems (IMS) will maintain the computerized automated production processes and equipment④. This high-performance automation represents the next generation of manufacturing systems. So far, many efforts have been made to investigate the architecture of the integrated environment for CIMS, since it is recognized as the key issue for this very comprehensive and sophisticated system.

Computer-integrated Manufacturing System

Computer-integrated manufacturing (CIM) is a strategic thrust, an operating philoso-

phy. Its objective is to achieve greater efficiencies within the entire business, across the whole cycle of product design, manufacturing and marketing. Currently, there exist islands of automation, such as computer-aided design, computer-aided manufacturing, flexible manufacturing systems, manufacturing resource planning, office automation, computerized marketed forecasting, and so on. Manual operations and paperwork systems as well as human decision making power link the islands together. CIM digitally connects these islands by providing fast, accurate, consistent data from decision-making process. CIM thereby can cut costs, enhance quality, reduce response time, and improve white-collar productivity. As a result, the competitive capacity of a product or company can gain an increased share in the international marketplace.

Selected from "Integrated Distributed Intelligent Systems in Manufacturing", Ming Rao, Qun Wang and Jianzhong Cha, Chapman & Hall, 1993.

New Words and Expressions

1. CIM=computer-integrated manufacturing **计算机集成制造**
2. intensify [in'tensifai] *v*. **增强，强化**
3. confront [kən'frʌnt] *v*. **使面对，遭遇**
4. TFP=total factor productivity **总的生产要素的生产率**
5. MLP=manufacturing labor productivity **制造劳动生产率**
6. IIM=integrated information management **集成信息管理**
7. simulation [simju'leiʃən] *n*. **仿真，模拟**
8. diagnosis [daiəg'nəusis] *n*. **诊断，识别**
9. academia [ækə'diːmjə] *n*. **学术界**
10. ICAD=intelligent computer-aided design **智能计算机辅助设计**
11. IMS=intelligent maintenance systems **智能维护系统**

Notes

① 参考译文：作为不断增强竞争的结果，已经加快了产品与系统开发的步伐，从而迫使制造商进入一个不断改善质量不只是竞争优势，而是一个生存问题的时代。

② 参考译文：计算机以数学模型为基础，辅助专家们进行数值分析、合成、仿真和图解，还能提供信息给他们作决策。

③ 参考译文：CIMS要做的是：将产品生命周期中的所有阶段，包括接受来自顾客的订单、编制生产计划、产品设计、制订工艺进度计划、数控编码、产品测试、销售、（售后）服务和生产管理，集成成为一个人工对操作过程干预最少的完整的信息处理系统。

④ 参考译文：在未来的工业公司中，智能计算机辅助设计（ICAD）将在人类专家的帮助下自动地生成和分析产品模型；CIMS技术将用来实现整个工厂的集成；而智能维护系统（IMS）将对计算机化的自动生产工艺和设备进行维护。

Exercises

1. After reading the text above, summarize the main ideas of it in oral English.

2. Answer the following questions according to the text.

① Describe the four stages of the development of industry automation.

② What technology or equipment can represent the stage of equipment-intensive industry?

③ Which stage does FMS belong to?

④ At which stage of the development of industry automation can we link the technologies such as CAD，CAM，FMS etc. together?

3. Translate the last paragraph into Chinese.
4. Put the following into Chinese by reference to the text.

a matter of M　　enterprise　　stand-alone　　architecture　　enhance

5. Put the following into English.

把注意力集中于　　市场　　计算机辅助工程

传感器　　柔性制造系统　　专家意见，专门技术

6. Translate the following sentences into English.

① CIM 系统不仅包含生产控制系统，还有生产计划和管理系统。

② CIM 系统旨在将工厂自动化与办公室自动化结合起来，构成整个公司的计算机网络。

Reading Material 28

Computer-Integrated Manufacturing (2)

Problem Definition in Manufacturing Industry

As we know，manufacturing industry has played an important role in the entire industrial development since the industrial revolution in the eighteenth century. However，with the advent of modern industry，especially the applications and development of computer techniques，the use of classical manufacturing techniques makes it difficult to fulfill people's expectations when facing current international competition. On the other hand，in implementing a computer-integrated manufacturing system，many problems may arise.

Too much information requires manipulation，especially with a newly implemented system. The stochastic occurrence of operational faults requires emergency handling in industrial manufacturing processes. For example，in the chemical industry，this issue is critical because of degraded operator efficiency and quality in handling emergencies，and due to the ever increasing degree of coupling between the system components and the response time requirements. We need to develop the automation of safe process startup and shutdown under irregular conditions to improve production efficiency and safety.

In many industrial companies，manufacturing processes are modern highly-computerized human-machine systems. Meanwhile，it is becoming increasingly difficult for operators to understand the many signals and information from display board，video，and computer screen. Owing to the importance of the operator's role in these systems，the quality of interaction between human operator and computers is crucial. Thus，development of an intelligent multimedia interface will better facilitate human interaction with complex realtime monitoring and control systems，Through the multimedia interface，computer systems can

communicate with operators via multimedia and modes, such as natural language, graphics, animation, and video presentation.

In a computer-integrated manufacturing systems, conflicting conclusions among production sections and different knowledge domains usually arise because different intelligent systems can make different decisions based on different criteria, even though the data that fire the rules are nearly the same. For example, when a disturbance occurs, the operation expert may change the operation state in a chemical process, but the control expert may wish to keep the operation state unchanged (due to the set-point control strategy). The lack of conflict-reasoning strategies has become the bottleneck in applications of integrated techniques. In addition, plant-wide optimization requires coordination among different production workshops and sections. Therefore, the integration of management, manufacturing and marketing information is a very important factor.

The amount of manufacturing information and management information is increasing fast. In manufacturing processes, a large amount of operational data and information from different sensors have to be collected and processed in time with useful knowledge and information acquired. Here, the key problems are (ⅰ) how to process so much data in time, and (ⅱ) how to extract effective knowledge which is useful for decision-making from new data, because all data cannot be stored in a computer due to excessive amount. Existing database technology and numerical computing methods may be used for operational data processing. However, how to effectively acquire knowledge from the data still remains difficult. Generally speaking, most of the problems involved in industrial manufacturing are usually illstructured, and difficult to formulate. Some important parameters cannot be measured online. Meanwhile, industrial manufacturing very often deals with uncertain and fuzzy information. In such process, mathematical modeling is not amenable, and purely algorithmic methods are difficult to use.

In manufacturing process, there is too much information needing to be represented. Some may be expressed explicitly in graphics or neural networks, rather than numerical or symbolic. So far, a few intelligent systems have been successful in integrating several expert systems and coupling symbolic reasoning with numerical computation. However, such integration still cannot satisfy the requirements of CIMS. The empirical data from process operations cannot be effectively processed with the existing expert systems or numerical models. Some process variables that affect process operation and product quality arc only partially understood. Sometimes, these variables are not directly measurable. In such a case, neural networks are an alternative to solve the problem. In fact, many complicated industrial problems cannot be solved by a single technique such as symbolic reasoning, numerical computation, or neural networks. A new methodology to integrate neural networks, symbolic reasoning, numerical computation, and computer graphics has to be developed to handle these problems. In addition, modern industrial processes are so complicated that no single tool can handle everything. For example, in an integrated operation environment, different existing software packages coded in computer languages such as C™, FORTOAN™ and

Pascal™, as well as written with commercial AI tools such as KEE™, OPS5™, G2™ and M. 1™, can be used together.

Nowadays, most process plants have highly automatic facilities. To stimulate industrial companies to utilize the newest manufacturing technology, we need to develop a system architecture that does not replace the existing factory facilities, operation environments, or information management systems in industrial plants, but effectively integrates and utilizes the knowledge and facilities that are currently available in companies.

To overcome the difficulties and solve the problems above, new techniques and methodologies should be introduced into the real industrial manufacturing environment.

Selected from "Integrated Distributed Intelligent Systems in Manufacturing", Ming Rao, Qun Wang and Jianzhong Cha, Chapman & Hall, 1993.

New Words and Expressions

1. manipulation [mənipju'leiʃən] *n*. 操纵，管理，处理
2. degraded [di'greidid] *a*. 被降级的，退化的
3. crucial ['kru:ʃəl] *a*. 关键的，决定性的，重要的
4. multimedia [mʌlti'mi:djə] *n*. 多媒体
5. facilitate [fə'siliteit] *v*. 使便利，促进
6. animation [æni'meiʃən] *n*. 动画
7. extract [iks'trækt] *v*. 提取
8. on-line ['ɔ:n,lain] *a*. 在线的，直接的；联机的，联用的
9. fuzzy ['fʌzi] *a*. 模糊的
10. neural ['njuərəl] *a*. 神经的，神经系统的
11. empirical [em'pirikəl] *a*. 经验的，以实验为基础的，经得起检验的

Unit 29 • Automatic Assembly (1)

The increasing need for finished goods in large quantities has, in the past, led engineers to search for and to develop new methods of production. Many individual developments in the various branches of manufacturing technology have been made and have allowed the increased production of improved finished goods at lower cost. One of the most important manufacturing processes is the assembly process. This process is required when two or more component parts are to be brought together to produce the finished product.

The early history of assembly process development is closely related to the history of the development of mass-production methods. Thus, the pioneers of mass production are also the pioneers of the modern assembly process. Their new ideas and concepts have brought significant improvements in the assembly methods employed in large-volume production.

However, although some branches of manufacturing engineering, such as metal cutting and metal forming processes, have recently been developing very rapidly, the technology of the basic assembly process has failed to keep pace. Table 29.1 shows that in the United States the percentage of the total labor force involved in the assembly process varies from about 20% for the manufacture of farm machinery to almost 60% for the manufacture of telephone and telegraph equipment①. Because of this, assembly costs often account for more than 50% of the total manufacturing costs. Statistical surveys show that these figures are increasing every year.

Table 29.1 Percentage of Production Workers Involved in Assembly

Industry	Percentage of workers involved in assembly
Motor vehicles	45.6
Aircraft	25.6
Telephone and telegraph	58.9
Farm machinery	20.1
Household refrigerators and freezers	32.0
Typewriters	35.9
Household cooking equipment	38.1
Motorcycles, bicycles, and parts	26.3

In the past few years, certain efforts have been made to reduce assembly costs by the application of automation and modern techniques, such as ultrasonic welding and diecasting. However, success has been very limited and many assembly operators are still using the same basic tools as those employed at the time of the Industrial Revolution.

Historical Development of the Assembly Process

In the early days of manufacturing technology, the complete assembly of a product was carried out by a single operator and usually, this operator also manufactured the individual component part of the assembly. Consequently, it was necessary for the operator to be an expert in all the various aspects of the work, and training a new operator was a long and expensive task. The scale of production was often limited by the availability of trained operators rather than by the demand for the product②.

In 1798, the United States needed a large supply of muskets and federal arsenals could not meet the demand. Because war with the French was imminent, it was also not possible to obtain additional supplies from Europe. However, Eli whitney, now recognized as one of the pioneers of mass production, offered to contract to make 10,000 muskets in 28 months. Although it took 10.5 years to complete the contract, Whitney's novel ideas on mass production had been successfully proved. The factory at New Haven, Connecticut, built specially for the manufacture of the muskets, contained machines for producing interchangeable parts. These machines reduced the skills required by the various operators and allowed significant increases in the rate of production. In an historic demonstration in 1801, Whitney surprised his distinguished visitors when he assembled musket locks after randomly selected parts from a heap.

The results of Eli Whitney's work brought about three primary developments in manufacturing methods. First, parts were manufactured on machines, resulting in a consistently higher quality than that of hand-made parts. These parts were now interchangeable and as a consequence assembly work was simplified. Second, the accuracy of the final product could be maintained at a higher standard, and third, production rates could be significantly increased.

Oliver Evans's conception of conveying materials from one place to another without manual effort led eventually to further developments in automation for assembly. In 1793, he used three types of conveyors in an automatic flour mill, which required only two operators. The first operator poured grain into a hopper and the second filled sacks with flour produced by the mill. All the intermediate operations were carried out automatically with conveyors carrying the material from operation to operation.

The next significant contribution to the development of assembly methods was made by Elihu Root. In 1849, Elihu Root joined the company that was producing Colt "six-shooters". Even though at that time the various operations of assembling the component parts were quite simple, he divided these operations into basic units that could be completed more quickly and with less chance of error. Root's division of operations gave rise to the concept "divide the work and multiply the output". Using this principle, assembly work was reduced to very basic operations and with only short periods of operator training, high efficiencies could be obtained.

Frederick Winslow Taylor was probably the first person to introduce the methods of time and motion study to manufacturing technology. The objective was to save the operator's time and energy by making sure that the work and all things associated to the work were placed in the best positions for carrying out the required tasks. Taylor also discovered that any worker has an optimum speed of working which, if exceeded, results in a reduction in overall performance.

Undoubtedly, the principal contributor to the development of production and assembly methods was Henry Ford. He described his principles of assembly in the following words:

"First, place the tools and then men in the sequence of the operations so that each part shall travel the least distance whilst in the process of finishing".

"Second, use work slides or some other form of carrier so that when a workman complete his operation he drops the part always in the same place which must always be the most convenient place to his hand and if possible have gravity carry the part to the next workman③".

"Third, use sliding assembly lines by which parts to be assembled are delivered at convenient intervals, spaced to make it easier to work on them".

These principles were gradually applied in the production of the Model T Ford automobile.

Selected from "Automatic Assembly", Geoffrey Boothroyd, Corrado Poli and Laurence E. Murch, Marcel Dekker, Inc., 1982.

New Words and Expressions

1. finished ['finiʃt] *a*. 完美的，精加工的，完工的
2. account for （总共）占；计算出；解释，说明
3. ultrasonic [ʌltrə'sɔnik] *n*.; *a*. 超声波，超声的
4. musket ['mʌskit] *a*. 火枪
5. arsenal ['aːsinl] *n*. 兵工厂，军械库，武器库
6. imminent ['iminənt] *a*. 危急的，急迫的
7. interchangeable [intə'tʃeindʒəbl] *a*. 可互换的，通用的
8. distinguished [dis'tiŋgwiʃt] *a*. 卓越的，杰出的
9. hopper ['hɔpə] *n*. 料斗，漏斗，贮斗
10. whilst=while [wailst] *conj*. 当……的时候

Notes

① 参考译文：表 29.1 给出了美国装配业所涉及的劳动力占总劳动力的百分比变化情况，从农业机械制造业中的约 20%到电话和电报设备制造业中的近 60%。

② 参考译文：生产规模不是受限于产品的需求，而是常常受限于是否能找到训练有素的操作者。

③ 参考译文：其次，使用工件滑道或其他形式的输送装置，以使得当一个工人完成其操作时能一直在其最顺手的同一位置放下零件，并且，如果可能的话，借助于重力将零件送至下一个工人。

Exercises

1. After reading the text above, write a summary of it.
2. Answer the following questions according to the text.

   ① What does "these figures" (in line 7, paragraph 3) refer to?

   ② According to the text, who was the first man to produce interchangeable parts in U. S.?

③ Who is the principal contributor to the development of production and assembly methods?

④ Describe Henry Ford's principle of assembly.

3. Translate the 7th and 8th paragraphs into Chinese.
4. Put the following into Chinese by reference to the text.

finished　machinery　die-casting　a large supply of　musket　arsenal　randomly　as a consequence

5. Put the following into English.

与……有关系，与……相关　大量生产，大量制造　和……齐步前进，和……并驾齐驱　工业革命　可互换的，通用的　导致，引起

6. Translate the following sentences into English.

① 在制造工业中采用自动装配可以提高生产效率，并能使产品维持在较高的质量水平。

② 采用自动装配工艺必须先保证零件具有可互换性。

Reading Material 29

Automatic Assembly (2)

The modern assembly line technique was first employed in the assembly of a flywheel magneto. In the original method, one operator assembled a magneto in 20 min. It was found that when the process was divided into 29 individual operations, carried out by separate operators working at assembly stations spaced along an assembly line, the total assembly time was reduced to 13 min 10 s. When the height of the assembly line was raised by 8 in., the time was reduced to 7 min. After further experiments were carried out to find the optimum speed of the assembly line conveyor, the time was reduced to 5 min, which was only one-fourth of the time taken by the original process of assembly. This result encouraged Henry Ford to utilize his system of assembly in other department of the factory, which were producing subassemblies for the car. Subsequently, this brought a continuous and rapidly increasing flow of subassemblies to the operators working on the main car assembly. It was found that the operators could not cope with increased flow, and it soon became clear that the main assembly would also have to be carried out on an assembly line. At first, the movement of the main assemblies was achieved simply by pulling them by a rope from station to station. However, even this development produced the amazing result of a reduction in the total time of assemble from 12 h 28 min to 5 h 50 min. Eventually, a power-driven endless conveyer was installed. It was flush with the floor and wide enough to accommodate a chassis. Space was provided for workers to either sit or stand while they carried out their operations and the conveyors moved at a speed of 6 ft/min past 45 separate workstations. With the introduction of this conveyor, the total assemble time was reduced to 93 min. Further improvements led to an even shorter overall assembly time and eventually, a production rate of one car every 10s of the working day was achieved.

The type of assembly operation dealt with above is usually referred to as operator assembly, and it is still the most widespread method of assembling mass-or large-batch-pro-

duced products. However, in certain cases, more refined methods of assembly have now emerged.

As a logical extension of the basic assembly line principle, methods of replacing operators by mechanical means of assembly have been devised. Here, it is usual to attempt to replace operators with automatic workheads where the tasks being performed were very simple and to retain the operators for tasks that would be uneconomical to mechanize. This method of assembly has rapid gained popularity for mass production and is usually referred to as automatic assembly. However, complete automation where the product is assembled completely by machine is essentially nonexistent.

Choice of Assembly Method

When considering the assembly of a product, a manufacturer has to take into account the many factors that affect the choice of assembly system. For a new product, the following considerations are generally important:

① Cost of assembly;

② Production rate required;

③ Availability of labor;

④ Market life of the product.

If an attempt is to be made to justify the automation of an existing operator assembly line, consideration has to be given to the redeployment of those operators who would become redundant. If labor is plentiful, the degree of automation depends on the reduction in cost of assembly and the increase in production rate brought about by the automation of the assembly line. However, it must be remembered that, in general, the capital investment in automatic machinery has to be amortized over the market life of the product unless the machinery may be adapted to assemble a new product. It is clear that if this is not the case and the market life of the product is short, automation is generally not justifiable.

Advantages of Automatic Assembly

Followings are some of the advantages of automation:

① Reduction in the cost of assembly;

② Increased productivity;

③ A more consistent product;

④ Removal of operators from hazardous operations.

A reduction in costs is often the main consideration and, except for the special circumstances listed above, it could be expected that automation would not be carried out if it was not expected to produce a reduction in cost.

Productivity in an advanced industrial society is an important measure of operating efficiency. Increased productivity, although not directly beneficial to manufacturer unless labor is scarce, is necessary to an expanding economy because it releases personnel for other tasks. It is clear that when put into effect, automation of assembly lines generally reduces the number of operators required and hence increases productivity.

Some of the assembly tasks that an operator can perform easily are extremely difficult to

duplicate on even the most sophisticated automatic workhead. An operator can often carry out a visual inspection of the part to be assembled, and parts that are obviously defective can be discarded. Sometimes a very elaborate inspection system is required to detect even the most obviously defective part. If an attempt is made to assemble a part that appears to be acceptable but is in fact defective, an operator, after unsuccessfully trying to complete the assembly, can reject the part very quickly without a significant loss in production. In automatic assembly, however, unless the part has been rejected by the feeding device, an automatic workhead will probably stop and time will then be wasted locating and eliminating the fault. If a part has only a minor defect, an operator may be able to complete the assembly, but the resulting product may not be completely satisfactory. It is often suggested that one of the advantages of automatic assembly is that it ensures a product of consistently high quality because the machine faults if the parts do not conform to the required specifications.

In some situations, assembly by operators would be hazardous due to high temperatures and the presence of toxic substances and other materials. Under these circumstances, assembly by mechanical means is obviously advantageous.

Selected from " Automatic Assembly", Geoffrey Boothroyd, Corrado Poli and Laurence E. Murch, Marcel Dekker, Inc., 1982.

New Words and Expressions

1. magneto [mæg'ni:təu] *n.* 永磁发电机
2. subassembly [sʌbə'sembli] *n.* 组件装配，部件装配
3. flush [flʌʃ] *a.* 同平面的，同高的（with）
4. chassis ['ʃæsi] *n.* 底盘，底（盘）架，底板［座］
5. redeployment [ˌri:di'plɔimənt] *n.* 调遣，调配
6. redundant [ri'dʌndənt] *a.* 多余的，过剩的，冗余的
7. amortize [ə'mɔ:taiz] *v.* 摊还，摊销，分期偿还；缓冲

Unit 30 • Lean Production, Agile Manufacturing and Mass Customization Production

Lean Production

Lean production, a new approach to production, emerged in the 1990s. It incorporates a number of the recent trends, with an emphasis on quality, flexibility, time reduction, and teamwork. This has led to a flattening of the organizational structure, with fewer levels of management①.

Lean production systems are so named because they use much less of certain resources than mass production systems use—less space, less inventory, and fewer workers—to produce a comparable amount of output. Lean production systems use a highly skilled workforce and flexible equipment. In effect, they incorporate advantages of both mass production (high volume, low unit cost) and craft production (variety and flexibility). And quality is higher than in mass production.

The skilled workers in lean production systems are more involved in maintaining and improving the system than their mass production counterparts. They are taught to stop production if they discover a defect, and to work with other employees to find and correct the cause of the defect so that it won't recur. This results in an increasing level of quality over time, and eliminates the need to inspect and rework at the end of the line.

Because lean production systems operate with lower amounts of inventory, additional emphasis is placed on anticipating when problems might occur before they arise, and avoiding those problems through careful planning②. Even so, problems still occur at times, and quick resolution is important. Workers participate in both the planning and correction stages. Technical experts are still used, but more as consultants rather than substitutes for workers. The focus is on designing a system (products and process) so that workers will be able to achieve high levels of quality and quantity.

Compared to workers in traditional systems, much more is expected of workers in lean production systems. They must be able to function in teams, playing active roles in operating and improving the system. Individual creativity is much less important than team success. Responsibilities also are much greater, which can lead to pressure and anxiety not present in traditional systems. Moreover, a flatter organizational structure means career paths are not as steep in lean production organizations. Workers tend to become generalists rather than specialists, another contrast to more traditional organizations.

Agile Manufacturing

Agility refers to the ability of an organization to respond quickly to demands or opportunities. It is a strategy that involves maintaining a flexible system that can quickly respond to changes in either the volume of demand or changes in product/service offerings. This is par-

ticularly important as organizations scramble to remain competitive and cope with increasingly shorter product life cycles and strive to achieve shorter development times for new or improved products and services③.

Agile manufacturing is a strategic approach to operations for competitive advantage that emphasizes the use of flexible operations to adapt and prosper in an environment of change. Agility involves a blending of several distinct competencies such as cost, quality, and reliability along with flexibility. Processing aspects of flexibility include quick equipment changeovers, scheduling and innovation. Product or service aspects include varying output volumes and product mix.

Successful agile manufacturing requires a careful planning to achieve a system that includes people, flexible equipment, and information technology. Reducing the time needed to perform work is one of the ways an organization can improve a key metric: productivity.

Mass Customization Production

Mass customization is a strategy of producing standardized goods or services, but incorporating some degree of customization in the final product or service. The increasing diversity of the customer requirements and the attraction of the mass production efficiency shift the major manufacturing mode from mass production to mass customization. Unlike mass production in which finished products need to be stocked in inventory and wait to serve customer's demands, mass customization considers fulfilling individual customer needs while maintaining near mass production efficiency. Unique information is provided by each customer so that the product can be tailored to his or her requirements. This mode of manufacturing requires the production system to be very flexible and its control system adaptive to the rapid changing customer demands.

Selected from "Operations management (8th edition) ", William J. Stevenson, McGraw-Hill/Irwin, 2005.

Selected from "An RFID-Based Distributed Control System for Mass Customization Manufacturing", Michael R. Liu, Q. L. Zhang, Lionel M. Ni, Mitchell M. Tseng, Springer-Verlag Berlin Heidelberg, 2004.

New Words and Expressions

1. lean production 精益生产
2. agile manufacturing 敏捷制造
3. mass customization production 大批量定制生产
4. recur [ri'kəː] *vi*. 重现，再来
5. consultant [kən'sʌltənt] *n*. 顾问，咨询者
6. substitute ['sʌbstitjuːt] *n*.; *v*. 代替者，替代品，替换，替代，代用品
7. generalist ['dʒenərəlist] *n*. 多面手，通才
8. agility [ə'dʒiliti] *n*. 敏捷
9. cope with 与……竞争，应付
10. strive [straiv] *v*. 努力，奋斗，力争，斗争
11. prosper ['prɔspə] *v*. 成功，使成功

Notes

① 参考译文：这就导致具有较少管理层次的扁平组织结构。

② 参考译文：因为精益生产系统是用较低库存量来运行，加之强调在问题形成前便预见何时会发生，以及通过仔细编制计划来规避那些问题。

③ 参考译文：当组织力图保持其竞争力和应付越来越短的产品寿命周期，以及力争缩短新的或改进的产品与服务的开发时间时，这就尤其重要。

Exercises

1. After reading the text above, write a summary of it.
2. Answer the following questions according to the text.

① What are the advantages of lean production systems over mass production systems?

② Compared to workers in traditional systems, what is expected of workers in lean production systems?

③ What does agility involve?

④ What is the difference between mass production and mass customization production?

3. Translate the 2nd paragraph into Chinese.
4. Put the following into Chinese by reference to the text.

mass production system　　teamwork　　substitute　　generalist　　agility

mass customization production

5. Put the following into English.

精益生产　　顾问，咨询者　　敏捷制造　　产品寿命周期　　生产力

6. Translate the following sentences into English.

① 精益生产是一种新的生产方法，它结合了近年来的生产趋势，并强调质量、灵活性、省时和协同工作。

② 敏捷制造可以对产品的数量需求或改变做出快速反应。

Reading Material 30

Virtual Manufacturing and Green Product Manufacturing

Virtual Manufacturing

Virtual Manufacturing (VM) is an integrated, synthetic manufacturing environment exercised to enhance all levels of decision and control in a manufacturing enterprise. VM can be described as a simulated model of the actual manufacturing setup which may or may not exist. It holds all the information relating to the process, the process control and management and product specific data. It is also possible to have part of the manufacturing plant be real and the other part virtual. Virtual manufacturing is the use of computer models and simulations of manufacturing processes to aid in design and production of manufactured products.

VM aims at providing an integrated environment for a number of isolated manufacturing technologies such as Computer Aided Design, Computer Aided Manufacturing, and Computer Aided Process Planning, thus allowing multiple users to concurrently carry out all or some of these functions without the need for being physically close to each other. For exam-

ple, a process planning engineer and a manufacturing engineer can evaluate and provide feedback to a product designer, who may be physically located in another state or country, at the same time as the design is being conceived.

Green Product Manufacturing

Environmentally conscious manufacturing requires environmental responsibility-greening-in all aspects associated with transforming raw materials into finished product within a factory. To achieve end-to-end green manufacturing, complex operations require a careful analysis and quantification of all input and output streams at three levels: the process step, the production line, and the factory. A system approach of this type includes efficient materials utilization, process monitoring, waste elimination or minimization, and materials recycling.

Green Product and Process Design

Product design decisions have a significant impact on the environment at all stages of the product life cycle. The designer can make a major contribution to the environmental impact of a product by including design for environment (DFE) criteria early in the design phase. The greenness of a product design is assessed using attribute hierarchies with 11 top-level attributes: subassembly reusability, label, internal joints, material variety, material identification, recycled content in material(s) used, chemical usage, additives, surface finishes, external joints, and hazard level of the material(s) used.

From a process perspective, the greenness of the subprocesses is based solely on the disposition of the by-products (output streams) that the processes generate. Fig. 30. 1 gives an overview of how to assess the greenness of a process step. A totally green process is an ecologically closed-loop system in which the outlet streams are completely recycled into the process steps. On the opposite side of the spectrum, the outlet streams are disposed of as toxic wastes into a landfill, the least desirable option from a green design perspective.

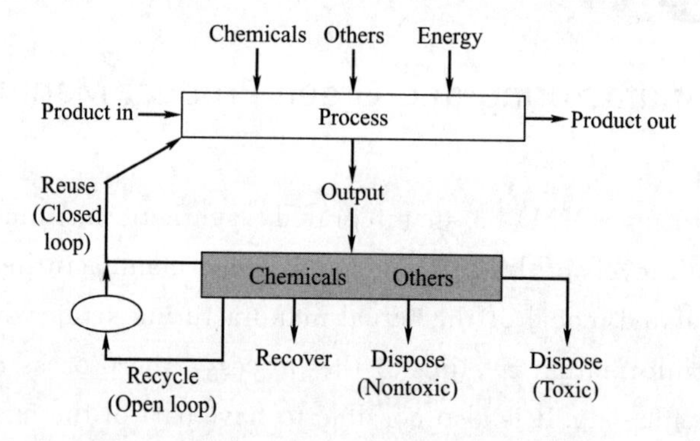

Fig. 30. 1 The material disposition of output streams determines the greenness of a process step

Green design of products and processes requires detailed information on subassemblies, materials used, and by-products generated by the processes. A major hurdle to performing a detailed evaluation of greenness is the lack of environmental information on many subassemblies in use today. If the information is readily available and can be accessed from local or on-line databases, designers can make a rapid green design assessment and refine it to

improve the greenness of product design and manufacturing processes. Wherever possible, the green design architecture incorporates such a feature, which is used to modify the product/process attributes where needed.

Minimizing Waste

Utilization and waste disposal of manufacturing process materials are becoming increasingly important issues for the manufacturing community. Increased public awareness and concern over the associated environmental considerations have raised everyone's level of consciousness and have brought these issues to the forefront of many agendas throughout the world. Regulations regarding manufacturing process materials utilization and waste disposal are evolving on state, federal, and international levels, along with more numerous, stringent, and complex regulations.

Selected from "Virtual Manufacturing: An Overview", Chetan Shukla, Michelle Vazquez and F. Frank Chen, 19th International Conference on Computers and Industrial Engineering, Vol. 31, No. 1/2, pp. 79-82, 1996, Elsevier Science Ltd.

Selected from "Green Product Manufacturing" David A. Dickinson, Clifton W. Draper, Manjini Saminathan, John E. Sohn and George Williams, AT & T Technical Journal, November/December 1995, 26-34.

New Words and Expressions

1. virtual manufacturing 虚拟制造
2. synthetic [sin'θetic] *a.* 人造的，综合的，假想的
3. manufacturing setup 生产装置
4. concurrently [kən'kʌrəntli] *ad.* 同时，兼，并行地
5. green product manufacturing 绿色产品制造
6. environmental conscious manufacturing 环境意识制造（绿色制造）
7. quantification [ˌkwɔntifi'keiʃ(ə)n] *n.* 量化
8. assess [ə'ses] *vt.* 估定，评定
9. attribute [ə'tribju(:)t] *n.* 属性，品质，特征
10. perspective [pə'spektiv] *n.* 观点，看法，（观察问题的）视角
11. ecological [ˌekə'lɔdʒikəl] *a.* 生态（学）的
12. landfill ['lændfil] *n.* 垃圾掩埋法，垃圾
13. hurdle ['hə:dl] *n.*; *v.* 障碍；克服（障碍）
14. stringent ['strindʒənt] *a.* 严格的，必须遵守的，严厉的

Appendix 1 • Vocabulary

abscissa	[æb'sisə]	n.	横坐标	3a
abrasive	[ə'breisiv]	n.	研磨剂，磨料	7a
abstraction	[æb'strækʃən]	n.	抽象概念，抽象；提取；抽取；分离	19a
ac=alternating current		n.	交流电	9a
academia	[ˌækə'diːmjə]	n.	学术界	28a
accelerometer	[ækˌselə'rəmitə]	n.	加速度计	15b
accommodate	[ə'kəmədeit]	v.	使适应	27b
account for			（总共）占；计算出；解释，说明	29a
ACFM=absolute cubic feet per minute			绝对立方英尺/分	13a
acrylic	[ə'krilik]	a.; n.	丙烯酸的；丙烯酸	8a
activation	[ˌæktə'veiʃən]	n.	活化（作用），活性（化）	8a
actuate	['æktjueit]	vt.	开［驱，起］动，使动作，操纵	15a
actuating signal			促动信号，执行信号	16a
actuator	['æktjuˌeitə]	n.	执行机构（元件），驱动器，促动器	20b
adaptability	[ədæptə'biliti]	n.	适应性	10a
adaptive	[ə'dæptiv]	a.	（自）适应的	6a
additive	['æditiv]	n.; a.	添加剂；增加的	2b
adjacent	[ə'dʒeisənt]	a.	相邻的，毗连的	2b
adjustment	[ə'dʒʌstmənt]	n.	调整，调节	16b
agile manufacturing			敏捷制造	30a
agility	[ə'dʒiliti]	n.	敏捷	30a
agitation	[ædʒi'teiʃən]	n.	（液体的）搅动，摇动	4a
algebraic	[ˌældʒi'breiik]	a.	代数学的，代数的	17a
algorithm	['ælgəriðəm]	n.	算法，计算程序	19a
align	[ə'lain]	v.	对中，（使，排）成一直线，调整	16a
alignment	[ə'lainmənt]	n.	对齐，校准，调整	23b
alkalinity	[ˌælkə'liniti]	n.	碱性，碱度	3a
Alloy Casting Institute (ACI)			合金铸造学会	1b
alphanumeric	[ˌælfənjuː'merik]	a.	字母数字的	24b
alternative	[ɔːl'təːnətiv]	a.; n.	可供替代的；可供选择的事物	19b
altar	['ɔːltə]	n.	祭坛，圣坛	10a
ambient	['æmbiənt]	a.; n.	周围的；周围环境	1a
ambient temperature			室温，环境温度	1a
amenable	[ə'miːnəbl]	a.	可处理的；经得起检验的	3b
American Iron and Steel Institute (AISI)			美国钢铁学会	1b
American Society for Testing and Materials (ASTM)			美国材料试验学会	1b
amortize	[ə'mɔːtaiz]	v.	摊还，摊销，分期偿还；缓冲	29b

amplification	[ˌæmplifiˈkeiʃən]	*n.*	放大，扩大	15b
amplitude	[ˈæmplitjuːd]	*n.*	（振，波）幅，幅度	22a
analog	[ˈænəlɔg]	*n.*	模拟（量），比拟	9a
and yet			可是，（然）而，但	15a
animation	[ˌæniˈmeiʃən]	*n.*	动画	28b
anneal	[əˈniːl]	*n.*; *v.*	退火	2a
anodic	[æˈnɔdik]	*a.*	阳极的	8b
anomaly	[əˈnɔməli]	*n.*	异常事物，反常现象	19a
antialiasing filter			去假频滤波器	21b
anti-jamming	[ˈænti ˈdʒæmiŋ]	*n.*	抗干扰	19b
anthropomorphic	[ˌænθrəpəuˈmɔːfik]	*a.*	拟人的，有人形的，类人的	5b
appendage	[əˈpendidʒ]	*n.*	附加物，附件	23a
approximately	[əprɔksiˈmətli]	*ad.*	大约，大致，近于	25b
apron	[ˈeiprən]	*n.*	（机床刀座下的）溜板箱，拖板箱	7b
aqueduct	[ˈækwiˌdʌkt]	*n.*	高架渠，渡槽	10a
arbitrary	[ˈɑːbitrəri]	*a.*	任意的	2a
armature	[ˈɑːmətjuə]	*n.*	电枢，（电机）转子，衔铁	9a
arsenal	[ˈɑːsinl]	*n.*	兵工厂，军械库，武器库	29a
as-cast			铸态的	1b
assembly	[əˈsembli]	*n.*; *v.*	装配	6a
assembly line			装配线	18b
assembly parts			装配［组装］件，组合零件	18b
assess	[əˈses]	*vt.*	估定，评定，评估	30b
assurance	[əˈʃuərəns]	*n.*	保证，担保	14a
attribute	[əˈtribju(ː)t]	*n.*	属性，品质，特征	30b
attribute *M* to *N*			把 *M* 赋于 *N*	15a
auditory cortex			听觉皮层	19a
augment	[ɔːgˈment]	*v.*; *n.*	增加［大、进］，扩张［大］；增加	18b
austenite	[ˈɔːstəˌnait]	*n.*	［冶］奥氏体	4a
austenitic	[ˌɔːstəˈnitik]	*a.*	奥氏体的	1b
austenitize	[ˈɔːstənətaiz]	*vt.*	［冶］奥氏体化，使产生奥氏体	4b
austere	[ɔsˈtiə]	*a.*	严峻的；苛刻的	18b
automated guided vehicles (AGV)			自动导向小车	27b
automation	[ɔːtəˈmeiʃən]	*n.*	自动化，自动操作	5b
availability	[əˌveiləˈbiliti]	*n.*	可用性，有效性，可得性	2a
axial	[ˈæksiəl]	*a.*	轴向的，轴线	3b
axiomatic	[ˌæksiəˈmætik]		公理的，自明的，自然的	4a
back-and-forth	[bæk ænd fɔːθ]	*ad.*	前后（运动），来回（运动）	13b
balance	[ˈbæləns]	*n.*	平衡，对称	22a
batch	[bætʃ]	*n.*	一次操作所需原料量，一次生产量，一批，批量	9b
bandwidth	[ˈbændwidθ]	*n.*	频带宽度，带宽，频宽	17a

bar code reader			条码阅读器	19b
be compatible with M			与M相容［相适应，一致，相似］，适合于M	27b
be coupled with			和……联合，结合	23a
be equipped with			装备着，安置着	25b
beam	[biːm]	*n.*	（横，天平）梁；光［射］线	3b
bearing	[ˈbɛəriŋ]	*n.*	轴承	5b
bend	[bend]	*v.*；*n.*	（使）弯［挠，折］曲；弯（管，头），弯曲（处）	3b
benign	[biˈnain]	*a.*	有益于健康的，（气候）温和的，良好的	4a
binary-coded		*a.*	二进制编码的	21b
bipod	[ˈbaipɔd]	*n.*	两脚台，两脚架	5b
bit	[bit]	*n.*	毕特（二进位数），位，存储信息容量单位	21b
biomedical		*a.*	生物医学的	12a
blacksmith	[ˈblæksmiθ]	*n.*	锻工，铁匠	6a
bladed impeller			装有叶片的叶轮	13a
blindfold	[ˈblaindfəuld]	*vt.*；*n.*；*a.*	蒙住……的眼睛，遮住……的视线；障眼物，遮眼物；看不清的，盲目的	16a
block valve			隔断阀，截断阀	10b
body-in-white assembly (BIWA)			白车身装配（是汽车行业中的特有术语，即未修饰喷漆前的汽车车身部件）	5b
bond	[bɔnd]	*v.*	粘结	7a
bonnet	[ˈbɔnit]	*n.*	阀帽，阀盖	10b
booth	[buːθ]	*n.*	亭，摊棚，摊位	25b
boring	[ˈbɔːriŋ]	*n.*	镗削，镗孔，扩孔	9b
braze	[breiz]	*vt.*	铜焊	1b
brittle	[ˈbritl]	*a.*	易碎的，脆性的	2b
broach	[brəutʃ]	*v.*	拉削	6b
buffer	[ˈbʌfə]	*n.*	缓冲器，减震器，保险杠，抛光轮	25a
bug	[bʌg]	*n.*	缺陷，困难，错误	27a
burnishing	[ˈbəːniʃiŋ]	*n.*	抛光	6b
calibration	[ˌkæliˈbreiʃən]	*n.*	校准	21a
capacity	[kəˈpæsitə]	*n.*	容量，生产量，生产力，能力	13b
carburetor	[kɑːbəˈretə(r)]	*n.*	汽化器	12a
carburization	[ˌkɑːbjuraiˈzeiʃən]	*n.*	渗碳，碳化	4b
carriage	[ˈkræridʒ]	*n.*	（机床的）拖板，机器的滑动部分	7b
cast ingot	[kæːstˈiŋgət]	*n.*	铸锭	1b
cast irons			铸铁	1a
cathode	[ˈkæθəud]	*n.*	阴极	8b
cause-and-effect relation			因果关系	17a
centrifugal	[senˈtrifjugəl]	*a.*	离心的	13b
CIM＝computer-integrated manufacturing			计算机集成制造	28a

chamber	['tʃeimbə]	n.	燃烧室，箱式，容器	4b
chassis	['ʃæsi]	n.	底盘，底（盘）架，底板［座］	29b
check valve			单向阀，止回阀	10b
chemical machining (CHM)			化学加工	8b
chisel	['tʃizl]	n.	凿子，錾子	3b
circuitry	['səːkitri]	n.	电路系统，（整机）电［网］路	21b
chromium	['krəumjəm]	n.	铬	2a
cladding	['klædiŋ]	n.	包层，覆盖，（金属）覆层	1b
closed-loop		n.	闭合回路［电路，环路］，闭环	9a
cluster	['klʌstə]	n.	群，组	26a
coefficient	[kəui'fiʃənt]	n.	［数］系数	12b
cognition	[kɔg'niʃən]	n.	认知，感知，认识	19a
coil	[kɔil]	n.	线圈	16b
cold-rolled	[kəuld rəuld]	a.	冷轧的	22b
collet	['kɔlit]	n.	弹性夹头，套筒，套爪	7b
combustible	[kəm'bʌstəbl]	a.; n.	易［可］燃的；可燃物	18b
commercialization	[kə'məːʃəlaizeiʃən]	n.	商业化，商品化	4a
commission	[kə'miʃən]	n.; v.	代理	6a
commutator	['kɔmjuteitə]	n.	换向器，整流器	17a
compact	['kɔmpækt]	a.	紧凑的，压缩的	24b
compactness		n.	紧密，紧密度，简洁，致密性	10a
comparator	['kɔmpəreitə]	n.	比较器［块，装置，电路］	21b
compatibility	[kəmˌpæti'biliti]	n.	相容性，可混性	2b
compliance	[kəm'plaiəns]	n.	依从，顺从，顺应，柔顺	5b
composite	['kɔmpəzit]	n.	合成，复合，复合材料	2b
compound rest			复式刀架，（车床）小刀架	7b
compression	[kəm'preʃən]	n.	压缩，压力	3a
compressor	[kəm'presə]	n.	压缩机	13a
Computer-Aided Manufacturing (CAM)			计算机辅助制造	6a
conceive	[kən'siːv]	v.	设想，想象	24a
concurrently	[kən'kʌrəntli]	ad.	同时，兼，并行地	30b
condense	[kən'dens]	v.	冷凝，凝结；浓［凝］缩	11b
conform to			与相符「一致」，符合，遵守	14b
confront	[kən'frʌnt]	v.	使面对，遭遇	28a
conjunction	[kən'dʒʌŋkʃən]	n.	连接，结合	4b
connecting rod			连杆，活塞杆	3b
consecutive	[kən'sekjutiv]	a.	连续的，连贯的	17b
consistency	[kən'sistənsi]	n.	一致性	25b
constituent	[kən'stitjuənt]	a.; n.	组成的，构成的；成分，组分	1a
consultant	[kən'sʌltənt]	n.	顾问，咨询者	30a
contaminate	[kən'tæmineit]	vt.	污染，弄脏	13a
continuous-data			连续数据	17b

continuous-path		*n.*	连续路径	9a
continuum	[kən'tinjuəm]	*n.*	连续（统一体），连续统	27a
contour	['kɔntuə]	*n.*	轮廓，周线，等高线，外形，造型	7a
convective	[kən'vektiv]	*a.*	对流的，传递性的	4b
cooling tower			冷却塔	11b
cope with		*v.*	与……竞争，应付	30a
corrosion	[kə'rəuʒən]	*n.*	腐蚀	1a
corrosive	[kə'rəusiv]	*a.*	腐蚀的，腐蚀性的	13a
counterboring	[kauntə'bɔːriŋ]	*n.*	镗阶梯孔，镗孔，锪平底孔	9b
couple M to [with] N			使M同N结［配、耦］合	15a
coupling	['kʌpliŋ]	*n.*	偶合，连接，轴接，联轴器（节）	22b
cracking	['krækiŋ]	*n.*	开裂，裂纹，裂缝	2a
crescent-shaped	['kresnt ʃeipd]	*a.*	新月形的，逐渐增加的	13b
creep	[kriːp]	*n.*	蠕变	3a
cross slide			横向滑板，横刀架，横拖板	7b
cross-section			横截面	3a
crucial	['kruːʃiəl]	*a.*	关键的，决定性的，重要的	28b
cryogenic	['kraiəu'dʒenik]	*a.*	低温的，深冷的	1b
crystalline	['kristəlain]	*a.*	结晶性的，晶状的	1a
crystallinity	['kristəlaiz]	*n.*	（结）晶性，结晶度	1a
cutoff frequency			截止频率	21a
cylinder	['silində]	*n.*	圆筒，圆柱体，汽缸，柱面	11a
cylindrical	[si'lindrik(ə)l]	*a.*	圆柱形的，圆筒形的，柱面的	25b
dam	[dæm]	*n.*	堤，坝	12a
damping	['dæmpiŋ]	*a.*	阻尼的，减振的，衰减的	22b
dashed	['dæʃt]	*a.*	虚（线）的	11a
dc=direct current		*n.*	直流电	9a
deadbeat controller			无差拍控制器	18a
decipher	[di'saifə]	*vt.*	解释，译解	24b
dedicated	['dedikeitid]	*a.*	专用的	27b
deficiency	[di'fiʃənsi]	*n.*	缺乏，不足，缺陷，欠缺	4b
deformable body			可变形物体	3b
deformation	[diːfɔː'meiʃən]	*n.*	变形	1a
defy	[di'fai]	*v.*；*n.*	挑战，挑衅	10a
degradation	[degrə'deiʃən]	*n.*	退化，降低，劣化	2b
degraded	[di'greidid]	*a.*	被降级的，退化的	28b
deliberation	[diˌlibə'reiʃən]	*n.*	考虑，熟虑，熟思	16a
delineate	[di'linieit]	*vt.*	描绘……的轮廓，描绘，描写	11a
delivery	[di'livəri]	*n.*	输出，交货	13a
deployment	[di'plɔimənt]	*n.*	（部队、资源或装备的）部署，调集	19a
derivative	[di'rivətiv]	*a.*；*n.*	导出［生］的，派生的；［数］导数，微商	18b
desired value			预期值	16a

destructive	[dis'trʌktiv]	*a.*	毁灭性的，破坏的，有害的	14a
deteriorate	[di'tiəriəreit]	*v.*	（使）恶化，退化	14a
deviation	[ˌdiːvi'eiʃən]	*n.*	偏离，偏移	3a
dexterity	[deks'teriti]	*n.*	灵巧，机敏	5b
dexterous	['dekstərəs]	*a.*	灵巧的，惯用右手的	5b
diagnosis	[ˌdaiəg'nəusis]	*n.*	诊断，识别	28a
diagonal	[dai'ægənl]	*a.*	对角（线）的，对顶（线）的	24b
diaphragm	['daiəfræm]	*n.*	隔膜，（电话等）振动膜	13b
diaphragm valve			隔膜阀	10b
dictate	[dik'teit]	*v.*	命令，支配	26a
dielectric	[daii'lektrik]	*a.*；*n.*	绝缘的，不导电的；电介质；绝缘体	8b
diesel	['diːzəl]	*n.*	柴油机	10a
digital signal processor（DSP）			数字信号处理器	18a
dimensional	[di'menʃənəl]	*a.*	线（维）度的，……维的	1a
diminish	[di'miniʃ]	*n.*	减小，缩小	3a
direct energy deposition			直接能量沉积	8a
disc	[disk]	*n.*	阀瓣，圆盘	10b
discernible	[di'səːnəbl，-'zəː-]	*a.*	可辨别得出的，可看出的	2a
discontinuity	['disˌkonti'nju(ː)iti]	*n.*	间断，不连续，中断	5a
discrete	[dis'kriːt]	*a.*	离散的，分立的，不连续的	17b
displacement	[dis'pleismənt]	*n.*	位移，平移，偏移	3b
disposal	[dis'pəuzəl]	*n.*	处理，清除，处理方法	9b
distinguished	[dis'tiŋgwiʃt]	*a.*	卓越的，杰出的	29a
distortion	[dis'tɔːʃən]	*n.*	变形，畸变	4a
distribute	[dis'tribju(ː)t]	*v.*	分布	6a
diverse	[dai'vəːs]	*a.*	不同的，（各种）各样的，多样的	7a
domain	[dəu'mein]	*n.*	领域，范围，域，区域	20a
domestication	[dəuˌmesti'keiʃən]	*n.*	家养，驯养	10a
dominant	['dɔminənt]	*a.*	占优势的，支配的，有统治权的	12b
dovetail	['dʌvteil]	*n.*	楔形榫头	7a
downstream	['daunstriːm]	*adv.*；*a.*	下游的	12b
dramatically	[drə'mætikəli]	*ad.*	显著地，引人注目地	15a
drilling	['driliŋ]	*n.*	钻孔，钻削	9b
ductile	['dʌktail]	*a.*	延性的，易变形的，可塑的，韧性的	1a
duct	[dʌkt]	*n.*	管，输送管，排泄管（指非圆形的）	12b
ductility	[dʌk'tiliti]	*n.*	延（展）性，韧性	2a
duplicate	['djuːplikeit]	*v.*	复制	25a
duplication	[ˌdjuːpli'keiʃən]	*n.*	复制品，成倍	7b
dynamics	[dai'næmiks]	*n.*	动力学，动态（特性）	3b
earthmoving	[3ːθˌmuːviŋ]	*a.*	大量掘土的，大量运土的	10a
ecological	[ˌekə'lɔdʒikəl]	*a.*	生态（学）的	30b
economizer	[iːkɔnəmaizə]	*n.*	省煤［油］器，废气预［节］热器	11b

economy of scale			规模经济	6a
eigenstructure			本征结构	18b
ejection	[i'dʒekʃən]	*n.*	发射，放射，弹射	15b
elastomer	[iː'læstəumə]	*n.*	弹性体，人造橡胶	13b
electrical conductor			导电体	22b
electrochemical machining (ECM)			电化学加工，电解加工	8b
electric discharge machining (EDM)			电火花加工	8b
electrolyte	[i'lektrəlait]	*n.*	电解质，电解液	8b
electroplating	[i'lektrəupleitiŋ]	*n.*	电镀，电镀术	8b
elongation	[ˌiːlɔŋ'geiʃən]	*n.*	延伸率，伸长	3a
embed	[im'bed]	*v.*	嵌入	20a
emergence	[i'məːdʒəns]	*n.*	出现，兴起	19b
empirical	[em'pirikəl]	*a.*	经验的，以实验为基础的，经得起检验的	28b
emulate	['emjuleit]	*v.*	模仿，仿真	18b
encoder	[in'kəudə]	*n.*	编码器	20a
encompass	[in'kʌmpəs]	*v.*	包含，包括	14a
endurance	[in'djurəns]	*n.*	持久性，耐久性	14a
energize	[enərdʒaiz]	*v.*	为……提供电力（或能量），使通电	19b
engine	['endʒin]	*n.*	发动机，引擎，火车头，机车	11b
enhance	[in'hɑːns]	*v.*	提高，增强	2b
environmental	[inˌvaiərən'mentl]	*a.*	环境的，环境产生的	12a
environmental conscious manufacturing			环境意识制造（绿色制造）	30b
esoteric	[ˌesəu'terik]	*a.*	深奥的，奥秘的	1a
essence	['esns]	*n.*	实质，本质	23a
etchant	['etʃənt]	*n.*	蚀刻剂，腐蚀剂	8b
evaluation	[iˌvælju'eiʃən]	*n.*	已取得数据的整理，求值，赋值，评估，估（测）算	26b
evaporator	[i'væpəreitə]	*n.*	蒸发［汽化］器	11b
evolve	[i'vɔlv]	*v.*	进化，发展，进展	6a
execution	[ˌeksi'kjuːʃən]	*n.*	执行，实行	25b
exemplify	[ig'zemplifai]	*vt.*	例证；作为……的例子	9a
exhaustive	[ig'zɔːstiv]	*a.*	无遗漏的，彻底的，详尽的	12a
expansion valve			膨胀［安全，调节］阀	11b
extended Kalman filtering			扩展卡尔曼滤波	18a
extender	[iks'tendə]	*n.*	填充剂，补充料	2b
extensive	[ik'stensiv]	*a.*	广泛的，大量的	19b
extract	[iks'trækt]	*v.*	提取	28b
extraneous	[eks'treinjəs]	*a.*	外来的，非必要的，无关的	17a
extrude	[eks'truːd]	*v.*	挤压	1b
fabricate	['fæbrikeit]	*vt.*	制造加工	1a
fabrication	[ˌfæbri'keiʃən]	*n.*	制造	2a
facilitate	[fə'siliteit]	*v.*	使便利，促进	28b

facing	['feisiŋ]	*n*.	车端面	6b
facsimile	[fæk'simili]	*n*.	复制品，传真	6b
failure	['feiljə]	*n*.	失效，破坏	3b
fallible	['fæləbl]	*a*.	易错的，可能犯错的	14b
family of parts			相似部件，零件族［组］	27a
fastener	['fɑːsnə]	*n*.	紧固件，接合件	4a
fatigue	[fə'tiːg]	*n*.	疲劳	3a
feasibility	[ˌfiːzə'biləiti]	*n*.	可行性	6b
federate	['fedərit]	*v*.	［使］联合，联盟	23b
feed rod			进给杆，光杆	7b
feedback	['fiːdbæk]	*n*.	反馈，回复	9a
feed-forward control			前馈控制	18a
feedrate	['fiːdreit]	*n*.	馈送率，进料速度，进给速度	15a
ferritic	[fə'ritik]	*a*.	铁素体的	1b
filament	['filəmənt]	*n*.	细丝，丝状物，单纤维	8a
filler	['filə]	*n*.	填充物，填料	2b
fillet	['filit]	*n*.	嵌条，(内) 圆角	5a
filter	['filtə]	*n*.; *v*.	滤波器，过滤器；筛选程序；过滤，用过滤法除去	18a
fingertip	['fiŋgətip]	*n*.	指尖，指套	5b
finished	['finiʃt]	*a*.	完美的，精加工的，完工的	29a
flange	[flændʒ]	*n*.	法兰 (盘)，凸缘	10b
flexible manufacturing			柔性制造	27a
flexibility	[ˌfleksə'biliti]	*n*.	柔性，灵活性，柔韧性	10a
flush	[flʌʃ]	*a*.	同平面的，同高的 (with)	29b
flux density			通量密度	21a
foresee	[fɔː'siː]	*vt*.	预见，预知	12a
forgeability	[ˌfɔːdʒə'biləti]	*n*.	可锻性	2a
foremost	['fɔːməust]	*a*.; *ad*.	最重要的；在最前	13a
fractional	['frækʃənl]	*n*.	分数的，分步的	9a
fracture	['fræktʃə]	*v*.; *n*.	(使) 断裂，(使) 破裂［碎］；断口［面］	5a
fume	[fjuːm]	*n*.	烟雾，烟	25b
furnace	['fəːnis]	*n*.	火炉，熔炉	16b
fused deposition modeling			熔融沉积造型	8a
fuzzy	['fʌzi]	*a*.	模糊的	28b
gage	[geidʒ]	*n*.	标准度量，计量器，量具	14a
gain	[gein]	*n*.	增益 (系数)，放大 (系数，率)，增量	21b
gain scheduling			增益调度，增益调节，增益规划	18a
gamut	['gæmət]	*n*.	范围	6b
gate valve			闸阀，滑门阀	10b
gear mesh			齿轮啮合	22b
generalist	['dʒenərəlist]	*n*.	多面手，通才	30a

generator	['dʒenəreitə]	*n.*	（蒸汽）发生器，发电［动］机	11b
generic	[dʒi'nerik]	*a.*	属的，类的，一般的，普通的	5b
geometric	[dʒiə'metrik]	*a.*	几何的，几何学的	5a
geometry	[dʒi'ɔmitri]	*n.*	几何学，几何图案	4b
globe valve			截止阀	10b
glossary	['glɔsəri]	*n.*	术语集，词集，专业词典	25a
governor	['gʌvənə]	*n.*	调节器，节制器	15a
grain	[grein]	*n.*	颗粒，晶粒	1a
grain boundary			晶界	1a
green product manufacturing			绿色产品制造	30b
grinder	['graində]	*n.*	磨床	25a
grinding	['graindiŋ]	*n.*	磨削	6b
grinding machine			磨床	7a
gyroscope	['gaiərəskəup]	*n.*	陀螺仪，回转仪	15b
gyro-stabilized		*a.*	陀螺稳定的	15b
handshake	['hændʃeik]	*n.*；*v.*	符号［信息］变换，握手信号	21b
harden	['hɑːdn]	*vt.*	使硬化，使变硬，淬火	4b
hardenability	[ˌhɑːdənə'biliti]	*n.*	可硬性；［冶］可淬性，淬透性	4b
hardness	['hɑːdnis]	*n.*	硬度	3b
harmonics	[hɑː'mɔniks]	*n.*	谐（调和）函数，谐波（频，音）	22a
harness	['hɑːnis]	*vt.*	利用（风等）作动力，治理，控制	10a
hazardous	['hæzədəs]	*a.*	危险的，有害的	25a
head	[hed]	*n.*	扬程	13b
headstock	['hedstɔk]	*n.*	主轴箱，床头箱	7b
heterogeneous	[ˌhetərə'dʒiːniəs]	*a.*	异类的，各种各样的	19a
hierarchical	[ˌhaiə'rɑːkikəl]	*a.*	分等级的，分级，分层，层次	18b
hierarchy	['haiərɑːki]	*n.*	分级结构，层次，等级制度	26a
honing	['həuniŋ]	*n.*	搪（珩）磨	7a
hopper	['hɔpə]	*n.*	料斗，漏斗，贮斗	29a
hurdle	['həːdl]	*n.*；*v.*	障碍；克服（障碍）	30b
HVAC＝Heating Ventilation and Air Conditioning			供暖通风与空气调节	20b
hybrid	['haibrid]	*a.*	混合的	23a
hydraulic instability			液压不稳定，水力不稳定	22a
hydrocarbon	['haidrou'kɑːbən]	*n.*	烃，碳氢化合物	13b
hydroelectric	['haidrəi'lektrik]	*a.*	水力电气的，水电的	12a
hysteresis	[ˌhistə'riːsis]	*n.*	滞后［现象，效应］，磁滞现象	21a
ICAD＝intelligent computer-aided design			智能计算机辅助设计	28a
ignite	[ig'nait]	*vt.*	点火，点燃	15b
IIM＝integrated information management			集成信息管理	28a
imminent	['iminənt]	*a.*	危急的，急迫的	29a
impact	['impækt]	*n.*	冲击，冲力，影响	3a
impart	[im'pɑːt]	*v.*	给予，产生	7a

impedance	[im'pi:dəns]	*n.*	阻抗	17a
impending	[im'pendiŋ]	*a.*	迫切的，即将发生的	26b
implementation	[ˌimplimen'teiʃən]	*n.*；*v.*	供给器具，装置，仪器；执行，实现	18b
implicit property			隐式属性	19a
impractical	[im'præktikəl]	*a.*	不（切）实际的，不现实的	14b
improve upon			（对……加以）改进（良），作出比……更好的东西	15a
input swing			输入摆幅	18a
IMS=intelligent maintenance systems			智能维护系统	28a
increment	['inkrimənt]	*n.*	增量，递增，增值	23b
incremental	[inkri'mentəl]	*a.*	增加的，增量的，递增的	24a
inertia	[i'nə:ʃjə]	*n.*	惯性，惯量，惰性，惰力，不活泼	18a
inevitably	[in'evitəbli]	*ad.*	不可避免地	17a
infiltration	[ˌinfil'treiʃən]	*n.*	渗入，渗透，渗透物	8a
inflated	[in'fleitid]	*a.*	膨胀的，夸张的，通货膨胀的	11a
inherent	[in'hiərənt]	*a.*	固有的，先天的，内在的	23b
inject	[in'dʒekt]	*n.*	注射，注入	12b
injection moulding			（塑料）注射成型	2b
inkjet printing and contour crafting			喷墨打印与轮廓加工	8a
innovation	[ˌinəu'veiʃən]	*n.*	改革，革新	23a
in-process		*a.*	在制，（加工，处理）过程中的	14a
inspection	[in'spekʃən]	*n.*	检验，检查	14a
instrumentation	[ˌinstrumen'teiʃən]	*n.*	仪表化	20a
insulation	[ˌinsə'leiʃən]	*n.*	绝缘，隔离	16b
integral	['intigrəl]	*a.*	整体的，完整的；总体的，总和的；必备的；积分的，累积的	18b
integration	[ˌinti'greiʃən]	*n.*	综合，结合，一体化，集成	23a
intensify	[in'tensifai]	*v.*	增强，强化	28a
interaction	[ˌintər'ækʃən]	*n.*	互相作用，互相影响，交互作用，互动	11a
interceptor	[intə'septə(r)]	*n.*	拦截战斗机	15b
interchangeability	[ˌintə(:)ˌtʃeindʒə'biliti]	*n.*	互换性，可交换性	20a
interchangeable	[intə'tʃeindʒəb(ə)l]	*a.*	可互换的，通用的	29a
interdependent variable			互依变量	20b
intergranular	[ˌintə'grænjulə]	*a.*	晶粒间的	4b
interior	[in'tiəriə(r)]	*a.*	内部的，本质的	16b
intermediate	[ˌintə'mi:diət]	*a.*	中间的，中级的	16b
intermittent	[ˌintə(:)'mitənt]	*a.*	间歇的，断断续续的	16a
internet of things (IoT)			物联网	19a
interpolate	[in'tə:pəuleit]	*v.*	插入，内插，插补	24b
intervention	[ˌintə(:)'venʃən]	*n.*	介入，干预	24a
invaluable	[in'væljuəbl]	*a.*	无价的，价值无法衡量的	14b
inventory	['invəntri]	*n.*	库存，存货（清单），报表	26a

iterate	['itəreit]	*v.*	迭代；重复，反复法	18b
keyseater	['kiːsiːtə]	*n.*	键槽铣床，铣键槽机	7a
keyway	['kiːwei]	*n.*	键（销）槽，销座，凹凸缝	7a
knurl	[nəːl]	*v. n.*	滚花，压花	7b
labor-intensive			劳动强度大的，劳动密集的	27a
ladder logic diagrams			梯形逻辑图	19b
ladder networks			梯形网络	20b
laminar	['læminə(r)]	*a.*	由薄片或层状体组成的，薄片状的	12b
laminated object manufacturing			叠层实体制造	8a
landfill	['lændfil]	*v.*；*n.*	垃圾掩埋法，垃圾	30b
landing gear			起落架	3b
lapping	['læpiŋ]	*n.*	研磨，精磨	7a
laser beam machining (LBM)			激光束加工	8b
lattice	['lætis]	*n.*	格子架，斜条结构	8a
lead	[liːd]	*n.*	（导，引）线，（电）线头	11b
lead screw			丝杆	7b
lead time			产品设计至实际投产间的时间，提前期，生产准备期	27b
leakage	['liːkidʒ]	*n.*	漏，泄漏，渗漏	11b
leaktight chamber			防漏腔	13b
lean production			精益生产	30a
least square			最小平方，最小二乘方	18a
lengthwise	['leŋθwaiz]	*a.*	纵向的	9a
limit switch			限位开关，行程开关	19b
linearity	[liniˈæriti]	*n.*	直线性	21b
line-of-sight			视线	20b
load	[ləud]	*n.*；*v.*	载［负］荷，（荷，负）载，加载	3b
longstanding		*a.*	长期间的，长期存在的	27b
loosen	['luːsn]	*v.*	解开，放松，松开	5a
magnetic relays			磁继电器	20b
magneto	[mægniːtəu]	*n.*	永磁发电机	29b
manifold	['mænifəuld]	*a.*；*vt.*	多的，多种多样的，许多种类的；增多，使……多样化	19a
manipulated variable			操纵量，调节变量	20b
manipulation	[mənipjuˈleiʃən]	*n.*	操纵，管理，处理	28b
manipulator	[məˈnipjuleitə]	*n.*	机械手，操作手，操纵器	25a
manual	['mænjuəl]	*a.*	手工的	6a
manufacturing setup			生产装置	30b
martempering	['mɑːtempəriŋ]	*n.*	［冶］等温淬火，热浴淬火，间歇淬火，马氏体回火	4a
martensite	['mɑːtənzait]	*n.*	［冶］马氏体	4a
martensitic	[mɑːtinˈzitik]	*a.*	马氏体的	1b

marvel	['mɑːvəl]	*n.*	奇异事物，奇迹	23a
maskant	['mɑskænt]	*n.*	保护层，掩蔽体	8b
mass customization production			大批量定制生产	30a
mass-production techniques			大批量生产技术	6a
master production schedule			总生产进度表	26b
matrix	['meitriks]	*n.*	基体，基质，矩阵	2b
maximum likelihood method			最大似然方法	18a
MCU＝monitor and control unit			监控设备，监控装置	26a
measurand	['meʒərənd]	*n.*	被测的物理量，测量变量	21a
mechanics	[mi'kæniks]	*n.*	力学，机械［构］学；机构，结构	3b
mechatronics	[mi'kætrɔniks]	*n.*	机械电子学，机电一体化	23a
metal-removal process			金属去除过程	6b
metallurgical	[ˌmetə'ləːdʒikəl]	*a.*	冶金的，冶金学的	14a
meteorology	[ˌmiːtjə'rɔlədʒi]	*n.*	气象学，气象状态	12a
methodology	[meθə'dɔlədʒi]	*n.*	方法学，方法论	12a
microprocessor		*n.*	微处理器，微型计算机，微信息处理机	21a
microsensor		*n.*	微型传感器	21a
microstructure	['maikrəu'strʌktʃə]	*n.*	显微结构	2a
milling	['miliŋ]	*n.*	铣削	6b
milling machine			铣床	7a
mimicking	['mimikiŋ]	*v.*	模仿（人的言行举止），（外表或行为举止）像，似	19a
misalignment	['misəlaimənt]	*n.*	不对中，不同心度，不平行度	22a
mislead	[mis'liːd]	*vt.*	使……误解，误导	2a
mist	[mist]	*n.*	烟雾，薄雾，模糊	4a
MLP＝manufacturing labor productivity			制造劳动生产率	28a
model reference adaptive control（MRAC）			模型参考自适应控制	18a
modulated signal			调制信号	20b
modulating	['mɔdjuleitiŋ]	*a.*	调制的	21a
module	['mɔdjuːl]	*n.*	模块，组件，模件	20a
monitor	['mɔnitə]	*n.*	监测	22a
monomer	['mɔnəmə]	*n.*	单（分子物）体，单基物	2b
mosaic	[mɔ'zeiik]	*n.*	镶嵌细工，马赛克	10a
moulding	['məuldiŋ]	*n.*	模塑（法），造型（法）	2b
multidisciplined	[ˌmʌlti'disiplind]	*a.*	多学科的	23a
multifunctional	[ˌmʌlti'fʌŋkʃənəl]	*a.*	多功能的	25a
multimedia	['mʌlti'miːdjə]	*n.*	多媒体	28b
multiplexer	['mʌltiˌpleksə]	*n.*	多路调制［转换］器，多路开关选择器	21b
musket	['mʌskit]	*a.*	火枪	29a
myriad	['miriəd]	*a.*	无数，无数的人［或物］	23a
nanocomposite	[ˌnænə(ʊ)'kəmpəzit]	*n.*	纳米复合材料	8a
necking	['nekiŋ]	*n.*	颈缩，形成细颈现象	3a

netting	[ˈnetiŋ]	*n.*	网，网状物，结网	4a
neural	[ˈnjuərəl]	*a.*	神经的，神经系统的	28b
niche	[niːʃ]	*n.*	适当的位置（场所）	8a
nickel	[ˈnikl]	*n.*	镍	2a
nomenclature	[nəˈmenklətʃər]	*n.*	命名法，专门术语	19a
nondestructive inspection			无损检测	14a
nonelectrical		*a.*	不用电的，非电的	21a
nonmetal	[ˈnɔnˈmetl]	*n.*	非金属	1a
nonrigorous	[ˈnɔnˈrigərəs]	*a.*	不严格的，不严密的	17a
nontraditional machining			特种加工	8b
notation	[nəuˈteiʃən]	*n.*	（符号）表示法，标志法	17a
notch	[nɔtʃ]	*n.*；*vt.*	槽口，凹口，刻痕；刻凹痕，开槽	5a
nuisance	[ˈnjuːsns]	*n.*	讨厌的人或东西，麻烦事，损害	5a
ntu	[nʌt]	*n.*	螺母，螺帽	10b
obsolete	[ˌɔbsəˈlit]	*a.*；*n.*；*vt.*	已过时的；被废弃的事物；淘汰，废弃	20b
offset	[ˈɔfset]	*n.*	偏移量，抵消，弥补	13b
on-line	[ˈɔːnˌlain]	*a.*	在线的，直接的；联机的，联用的	28b
opaque	[əuˈpeik]	*a.*	不透明的	1a
open-loop		*n.*	开口［非闭合］回路，开环	9a
optical	[ˈɔptikəl]	*a.*	光学的，光导的	20a
ordinate	[ˈɔːdinit]	*n.*	纵坐标	3a
orientation	[ˌɔ(ː)rienˈteiʃən]	*n.*	定向，定位，排列方向	1a
origination	[əˌridʒəneitiv]	*n.*	产生，出现，发明	24a
orthodox	[ˈɔːθədɔks]	*a.*	传统的，正统的，惯常的，普通的	15a
oscillate	[ˈɔsileit]	*vi.*	振荡，摆动	16b
overall gain			总增益	17a
oversee	[ˈəuvəˈsiː]	*v.*	管理，照料	26a
oxidation	[ɔksiˈdeiʃən]	*n.*	氧化	1b
packing gland			填料压盖，密封压盖	10b
pallet	[ˈpælit]	*n.*	板台，滑板，托板，托盘	9b
parabolic	[ˌpærəˈbɔlik]	*a.*	抛物线的，抛物面的	9a
paramount	[ˈpærəmaunt]	*a.*	最重要的，头等的，最高的	14a
particle	[ˈpɑrtikəl]	*n.*	［数，物］质点，粒子	3b
parting	[ˈpɑːtiŋ]	*n.*	切断，分离	7b
pause	[pɔːz]	*v.*；*n.*	暂停，停顿	19a
payload	[ˈpeiˌləud]	*n.*	净载重量，有效载荷	5b
peening	[ˈpiːniŋ]	*v.*	［冶］锤击硬化，喷丸硬化	4a
penetrate	[ˈpenətreit]	*v.*	穿过，进入，渗透	19b
perception	[pərˈsepʃən]	*n.*	知觉，感知，洞察力，看法，见解	19a
perforate	[ˈpəːfəreit]	*v.*	冲孔，穿孔	24a
performance	[pəˈfɔːməns]	*n.*	性能，特性；行为，操作，工况，绩效	4a
periodic	[piəriˈɔdik]	*a.*	周期的，定期的，循环的	27a

perspective	[pəˈspektiv]	*n.*	观点，看法，（观察问题的）视角	30b
pertain	[pə(:)ˈtein]	*vi.*	从属于，适合	9b
petrochemical	[ˌpetrəuˈkemikəl]	*a.*；*n.*	石化的；石化产品	13a
pigment	[ˈpigmənt]	*n.*	染料，色素	2b
piston	[ˈpistən]	*n.*	[机] 活塞，瓣	5b
pitch	[pitʃ]	*n.*	俯仰，倾斜，齿距，螺距，节距	25b
planer	[ˈpleinə]	*n.*	龙门刨床	7a
plasma	[ˈplæzmə]	*n.*	[物] 等离子区，等离子体	4b
plasticizer	[ˈplæstisaizə]	*n.*	增塑剂，柔韧剂	2b
plunger	[ˈplʌndʒə]	*n.*	柱塞	13b
pneumatic	[nju(:)ˈmætik]	*a.*	气动的，空气的	9b
point-to-point		*a.*	点至点的，点位控制，定向的	9a
polaris/poseidon fleet Ballistic Missile (FBM)			北极星式/海神式舰载弹道导弹	15b
polarization	[ˌpəuləraiˈzeiʃən]	*n.*	极化，偏振	21a
polymer	[ˈpɔlimə]	*n.*	聚合物，聚合材料	2b
polymer-derived ceramifiable monomers			先驱体转化法制备陶瓷	8a
polymerization	[ˌpɔliməraiˈzeiʃən]	*n.*	聚合作用，聚合反应	2b
position	[pəˈziʃən]	*v.*	确定……的位置，定位	24a
positioning	[pəˈziʃəniŋ]	*n.*	定位，固定位置	9a
post-processing			后处理，后置处理	20a
potentiometer	[pəˌtenʃiˈɔmitə]	*n.*	电位计，电势计	17b
powder bed fusion			粉末床融合	8a
powder-metallurgy	[ˈpaudəmeˈtælədʒi]	*n.*	粉末冶金学	1b
power plant			[动力] 发电厂，动力设备 [装置]	11b
precede	[pri(:)ˈsi:d]	*v.*	在……之前，优于，较……优先	13a
precise	[priˈsais]	*a.*	精确的	16b
predecessor	[ˈpri:disesə]	*n.*	（被代替的）原有（事）物，前任	26a
predetermined	[ˌpridiˈtə:mind]	*a.*	先已决定的，预先确定的	16b
predictive maintenance			预测性维修	22a
preheat	[ˈpri:ˈhi:t]	*v.*	预热	2a
prehistoric	[ˈpri:hisˈtɔrik]	*a.*	史前的，很久以前的	10a
preload	[ˈpri:ˈləud]	*vi.*	预加载，预装入	5b
premanufacture	[pri:məˈraitəl]	*n.*	预制造	14a
pre-processing			前处理，前置处理	20a
prevail	[priˈveil]	*vi.*	占优势，经常发生	3b
prismatic	[prizˈmætik]	*a.*	棱形的，棱柱的，棱镜的	5b
proficient	[prəˈfiʃənt]	*a.*；*n.*	精通的，熟练的；高手，专家	15b
profile	[ˈprəufail]	*n.*	轮廓，断面	24b
profiling	[ˈprəufailiŋ]	*n.*	试探，探测，检验，测深	7a
profound	[prəˈfaund]	*a.*	深刻的，意义深远的，渊博的	12b
programmable logic controller (PLC)			可编程控制器	19b
proliferation	[prəˌlifəˈreiʃən]	*n.*	激增，涌现	19a

prolong	[prəˈlɔːŋ]	*vt.*	延长，拖延	16b
propel	[prəˈpel]	*vt.*	推进	15b
propeller	[prəˈpelə]	*n.*	推进器，螺旋桨	12a
propulsive	[prəuˈpʌlsiv]	*a.*	推进的，有推进力的	12a
prosper	[ˈprɔspə]	*v.*	成功，使成功	30a
prototype	[ˈprəutətaip]	*n.*	样机，原型	5b
proximity	[prɑːkˈsiməti]	*n.*	接近，邻近，靠近	19b
proximity switch			接近开关	19b
pulley	[ˈpuli]	*n.*	滑车，滑轮	10a
punching	[ˈpʌntʃiŋ]	*n.*	冲压	6b
quantification	[ˌkwɔntifiˈkeiʃ(ə)n]	*n.*	量化	30b
quartz	[kwɔːts]	*n.*	石英	2b
quench	[kwentʃ]	*vt.*	把……淬火，使骤冷，使淬硬	4a
quenchant	[ˈkwentʃənt]	*n.*	淬火剂	4a
quill	[kwil]	*n.*	钻杆，套管轴，空心轴，滚针，小镗杆	7b
R and D=research and development			研究与发展，研制与试验，研究与开发（研发）	26a
radiation	[ˌreidiˈeiʃən]	*n.*	［物］发光，发热，辐射，放射物，辐射能	4b
radiator	[ˈreidieitə]	*n.*	散热器，电暖炉，辐射体	5a
ram	[ræm]	*n.*	（牛头刨）滑枕	7a
randomly		*ad.*	随机地，随便地，未加计划	12b
ratio	[ˈreiʃiəu]	*n.*	比率，比例	16b
ready bank			现成数据库，就绪数据库	26b
realization	[ˌriəlaiˈzeiʃən]	*n.*	实现	27a
ream	[riːm]	*vt.*	（用铰刀）铰孔	2a
reassess	[ˈriːəˈses]	*v.*	重估，再评价	23a
reciprocate	[riˈsiprəkeit]	*v.*	（使）往复（运动）	7a
recondition	[ˈriːkənˈdiʃən]	*vt.*	修理［复，整，补］，检［翻，整］修	14b
reconfigure	[ˌriːkənˈfigə(r)]	*v.*	重新配置［组合］	27a
recur	[riˈkəː]	*vi.*	重现，再来	30a
recursive	[riˈkəːsiv]	*a.*	递归的，循环的	18a
redeployment	[ridiˈplɔimənt]	*n.*	调遣，调配	29b
redundant	[riˈdʌndənt]	*a.*	多余的，过剩的，冗余的	29b
reemphasize	[riˈemfəsaiz]	*v.*	反复强调	23a
refrigerant	[riˈfridʒərənt]	*a.*; *n.*	制冷的；制冷剂	12a
refrigeration	[rifridʒəˈreiʃən]	*n.*	冷藏，制冷，冷却	13b
refrigerator	[riˈfridʒəreitə]	*n.*	制冷器［机］，冷气机，（电）冰箱	11b
relief valve			安全［减压，卸压，保险］阀	10b
renaissance	[rəˈneisəns]	*n.*	文艺复兴	20a
repetitive	[riˈpetitiv]	*a.*	重复的，反复性的	14b
reset	[ˈriːset]	*vt.*; *n.*	重新调整，重新安装	14b
resolution	[ˌrezəˈluːʃən]	*n.*	分辨率，清晰度；决定，解决	8a

resonant	['rezənənt]	*a.*	共振的，引起共鸣的	22b
resultant	[ri'zʌltənt]	*n.*；*a.*	合力；合成的，总的	3b
retract	[ri'trækt]	*v.*	缩回，收回	24b
retrieve	[ri'triːv]	*v.*	更正，恢复，取回，检索	6b
revolute	['revəl(j)uːt]	*a.*；*vt.*	旋转的，向后卷的，外卷的；旋转	5b
rigid body			刚体，刚性体	3b
rigor	['rigə]	*n.*	艰苦，酷烈，严格	27b
ripple	['ripl]	*n.*	波纹，涟漪	23a
roll	[rəul]	*v.*	轧制	1b
rotary screw compressor			回转式螺旋压缩机	13a
roving	['rəuviŋ]	*a.*	流动的	14b
rubbing	['rʌbiŋ]	*n.*	摩擦，研磨	13a
rudder	['rʌdə]	*n.*	方向舵，舵	16a
rudimentary	[ruːdi'mentəri]	*a.*	根本的，未发展的	12a
rugged	['rʌgid]	*a.*	结实的，坚固的	5a
saddle	['sædl]	*n.*	鞍状结构，滑动座架，滑座，床鞍	9a
sample	['sæmpl]	*n.*	样品，试样，取样	3b
sample-and-hold			样品保持	21b
saturation	[ˌsætʃə'reiʃən]	*n.*	饱和	17b
scaffold	['skæfəuld]	*n.*	支架	8a
scalability	[ˌskeilə'biliti]	*n.*	可伸缩性，可缩放性，可量测性	5b
scenario	[sə'næriəu]	*n.*	设想，方案，预测	19a
schematic	[ski'mætik]	*a.*	示意性的	11a
scrap	[skræp]	*v.*	扔弃，敲碎，拆毁，报废	25b
screw pump			螺杆泵，螺旋泵	13b
screwdriver	['skruːdraivə]	*n.*	螺丝起子，改锥，旋凿，螺丝刀	25a
seal	[siːl]	*v.*；*n.*	密封	11b
seat	[siːt]	*n.*	阀座	10b
self-priming	[self'praimiŋ]	*a.*	自吸的	13b
self-tuning regulator（STR）			自校正调节器	18a
semiautomatic	['semiˌɔːtə'mætik]	*a.*	半自动的	6a
sense	[sens]	*vt.*；*n.*	检测，断定，读出，感知	15a
sensor	['sensə]	*n.*	传感器，检测器	21a
serendipity	[ˌserən'dipiti]	*n.*	运气，善于发掘新奇事物的才能	23a
serial-parallel			串行-并行	21b
serration	[se'reiʃən]	*n.*	齿面，锯齿（形，构造）	7a
servo	['səːvəu]	*n.*	伺服，伺服系统，随动装置	5b
servo mechanism			伺服机构，伺服机械	15a
servo valve			伺服阀，继动阀	9a
shaft	[ʃɑːft]	*n.*	传动轴，旋转轴	3b
shaper	['ʃeipə]	*n.*	牛头刨床，插床	7a
shear	[ʃiə]	*n.*	剪切，切应变	3a

shell	[ʃel]	*n.*	壳，管壳，外壳；套管，罩	4a
silicon oxycarbide			碳氧化硅	8a
simultaneously	[ˌsiməl'teiniəsly]	*ad.*	同时，同时发生	23b
simulation	[ˌsimju'leiʃən]	*n.*	仿真，模拟	28a
sintering	['sintəriŋ]	*n.*	烧结，熔结	8a
sketch	[skætʃ]	*n.*；*v.*	草图，示意图；画草图，画示意图	22b
slant	[slɑːnt]	*v.*；*n.*	倾斜	9b
sleeve	[sliːv]	*n.*	套（筒、管、轴、环），轴套	15a
sluggish	['slʌgiʃ]	*a.*	惰性的；黏滞的；不活泼的；小［低］灵敏度的	18a
smoothing filter			平滑滤波器	21b
Society of Automotive Engineers（SAE）			美国汽车工程师学会	1b
sodium	['səudjəm]	*n.*	［化］钠	13b
solder	['səldə]	*vt.*	钎焊	1b
solenoid	[sɑːlənɔid]	*n.*	螺线管	19b
sophisticated	[sə'fistikeitid]	*a.*	复杂的	16b
sophistication	[səˌfisti'keiʃən]	*n.*	复杂化，完善（化），采用先进技术	16a
space heating			环流供暖	11b
specific weight			比重	13b
specimen	['spesimən]	*n.*	试件，试样	3a
spectrographic	['spektrəuˌgrɑːfik]	*a.*	摄谱仪的，光谱的	14a
spectrum	['spektrəm]	*a.*	范围，系列，领域，谱（如：光谱、频谱等）	26a
spectrum changing			频谱改变	21b
spherical	['sferikəl]	*a.*	球形的，球状的	25b
spill	[spil]	*v.*；*n.*	溢出，流出	10a
spin	[spin]	*v.*	旋压	1b
spotting	['spɔtiŋ]	*n.*	钻中心孔，找正	9b
spring	[spriŋ]	*n.*	弹簧，弹性，弹力	4a
spur	[spəː]	*v.*；*n.*	刺激，激励，鼓励，推动	10a
stabilizer	['steibilaizə]	*n.*	稳定剂	2b
stand-alone		*a.*	可独立应用的，单独的	27a
statics	['stætiks]	*n.*	静力学，静（止状）态	3b
steering wheel		*n.*	舵轮，转向［方向，驾驶］盘	16a
steam drum			蒸汽锅筒，上汽锅	11b
stem	[stem]	*n.*	阀杆，杆	10b
stepwise	['stepwaiz]	*a.*；*ad.*	逐步（的），逐渐（的），分段的，阶式的	21b
stereolithography			激光立体印刷术，光固化立体造型	8a
stochastic	[stəu'kæstik]	*a.*	随机的，机遇的，不确定的，概率性的	18a
stochastic approximation			随机逼近（法）	18a
straight-cut		*n.*	纵向切削	9a
strain	[strein]	*n.*	应变	3b
strength	[streŋθ]	*n.*	强度（极限），浓度，力（量，气）	3b
stress	[stres]	*n.*	应力	3b

stress concentration			应力集中	5a
stringent	['strindʒənt]	*a.*	严格的，必须遵守的，严厉的	30b
strive	[straiv]	*v.*	努力，奋斗，力争，斗争	30a
strut	[strʌt]	*n.*	支柱；抗压构件	3b
stuffing	['stʌfiŋ]	*n.*	填料，填充剂，加脂	9b
stuffing box			填（料）函，填料箱［盒］	10b
subassembly	['sʌbə'sembli]	*n.*	组件装配，部件装配	29b
subdivide	['sʌbdi'vaid]	*v.*	［把……］再分，［把……］细分	13a
subdivision	['sʌbdiˌviʒən]	*n.*	细分，细分度	24b
suboptimal	['sʌb'ɔptiməl]	*a.*	次优的，未达最佳标准的，不最适宜的	18a
subsequent	['sʌbsikwənt]	*a.*	后来的，接下去的	11a
subsequently	['sʌbsikwəntli]	*ad.*	其次，其后，接着	24a
substitute	['sʌbstitjuːt]	*n.*；*v.*	代替者，替代品，替换，替代，代用品	30a
substrate	['sʌbstreit]	*n.*	基质，基材，底层，基底，底物	8a
summing circuit			加法［求和］电路	20b
superiority	[sju(ː)piəri'ɔriti]	*n.*	优越性	13a
supervisory	[ˌsjuːpə'vaizəri]	*a.*	监督的，管理的	20a
synchro	['siŋkrəu]	*n.*	自整角机，同步机	17b
synchronize	['siŋkrənaiz]	*v.*	同步化，（使）同步，（使）同时（发生）	9a
synergistic	['sinədʒistik]	*a.*	协同的，协作的，叠加的，复合的	23a
synthesize	['sinθisaiz]	*v.*	（人工）合成，制造；综合（处理）	18b
synthetic	[sin'θetic]	*a.*	人造的，综合的，假想的	30b
tabulating	['tæbjuleitiŋ]	*n.*	用表格表示	9b
tachometer	[tæ'kɔmitə]	*n.*	转速计，流速计	9a
tailstock	['teilstɔk]	*n.*	尾架，尾座，顶尖座	7b
tangential	[tæn'dʒenʃ(ə)l]	*a.*	切线的，肤浅的，略为触及的	6a
tar	[tɑː]	*n.*	焦油	13a
tenfold	['tenfəuld]	*a.*；*ad.*	十倍的［地］	24b
tensile	['tensail]	*a.*	受拉的，拉伸的	3a
terminology	[ˌtəːmi'nɔlədʒi]	*n.*	专门名词，术语，词汇，术语学	18b
texture	['tekstʃə]	*n.*	纹理，［织物的］密度，［材料的］结构	14a
TFP=total factor productivity			总的生产要素的生产率	28a
thermocouple	['θəːmoukʌpl]	*n.*	热电偶	19b
thermodynamic	['θəːməudai'næmik]	*a.*	热力学的，使用热动力的	13a
thermodynamics	['θəːməudai'næmiks]	*n.*	［物］热力学	11a
thermoplastic	[ˌθəːmə'plæstik]	*a.*；*n.*	热塑性的；热塑性塑料	2b
thermosetting	[ˌθəːməu'setiŋ]	*a.*	热固（凝）的	2b
thermosetting polymer			热固性塑料	2b
thermostat	['θəːməstæt]	*n.*	恒温器，温度自动调节器	15b
thread	[θred]	*n.*	螺纹［齿，丝，线］，线（状物）	10b
three-dimensional	[θriː'dimenʃənəl]	*a.*	三维的，3D 的	24b
three-way valve			三通阀	10b
throttle	['θrɔtl]	*v.*	节流［气］，用（节流阀）调节	
		n.	节流［气］阀，风［油，主气］门	10b

time-consuming	['taimkən,sju:miŋ]	a.	费时（间）的，拖延时间的	27b
time-invariant			定常的	17b
time-lag			延时，落后，时滞	3a
time-shared		a.	分时的	21b
time-varying			时变的	17b
tolerance	['tɔlərəns]	n.；vt.	公差，容许量 给（机器部件等）规定公差	5a
tonnage	['tʌnidʒ]	n.	（总）吨位	1b
tool post			刀架，刀座	7b
torsion	['tɔ:ʃən]	n.	扭转，转矩	3a
toxic	['tɔksik]	a.	有毒的，中毒的	25b
trace back(...)to...			（把……）追溯到……	20a
trade-off		n.	折中（办法，方案），权衡，综合	27b
traditional	[trə'diʃən(ə)l]	a.	传统的，惯例的	27b
trajectory	['trædʒiktəri]	n.	轨道，弹道，轨迹	15b
transducer	[træz'dju:sə]	n.	变换器，换能器，传感器	20a
transfer function			转移函数	18a
transfer line			组合机床自动线	27a
transformation	[,trænsfə'meiʃən]	n.	变形，转化，转换	4a
transient	['trænziənt]	a.	瞬时的，瞬变的	26b
transient response			瞬态响应	17a
transition	[træn'ziʒən]	n.	转变，转换，变迁，过渡时期	27a
transmitter	[træns'mitə]	n.	发送器，发射机	20b
traverse	['trævə(:)s]	v.	横行，横向，横动	24b
trepanning	[tri'pæniŋ]	n.	穿孔	6b
troubleshooter	['trʌb(ə)lʃu:tə(r)]	n.	故障检修工	27a
troubleshooting	['trʌb(ə)lʃu:tiŋ]	n.	发现并修理故障，排除故障	27a
turbine	['tə:bin]	n.	涡轮（机），透平（机），汽轮机	11a
turbulence	['tə:bjuləns]	n.	骚乱，动荡，湍流或紊乱	12b
turbulent	['tə:bjulənt]	a.	湍动的，湍流的	12b
turning	['tə:niŋ]	n.	车削	6b
turn out to			原来是	17a
two-edged sword			双刃剑，两面性	17a
two-fold			两倍，两方面	4b
typesetter	['taip,setə]	n.	字母打印机	24a
ubiquitous	[ju:'bikwitəs]	a.	处处存在的，普遍存在的	1a
ultrasonic	['ʌltrə'sɔnik]	n.；a.	超声波，超声的	29a
ultrasonic machining（USM）			超声加工，超声波加工	8b
underlying rules			潜规则，潜在的规律	19a
unimation	[ju:nimə(r)]	n.	通用机械手（一种机器人的商品名）	25a
unmodulated	[ʌn'mɔdjuleitid]	a.	未调制的，非调制的，未经调制的	17b
unprecedented	[ʌn'presidentid]	a.	前所未有的，空前的，没有先例的	20b
unsurpassable	[,ʌnsə'pɑ:səbl]	a.	无法超越的	9b
Utopia	[ju:'təupiə]	n.	乌托邦，理想的完美境界	23b

validate	[ˈvælideit]	*vt.*	使生效，证实	4a
valve	[vælv]	*n.*	阀，阀门	10b
vanadium	[vəˈneidiəm]	*n.*	钒	2a
vane	[vein]	*n.*	叶片	13a
variable	[ˈveriəbl]	*a.*；*n.*	变化的，可变的；变量	20b
variation	[ˌveriˈeiʃən]	*n.*	变化，变动，演变	20b
ventilate	[ˈventileit]	*v.*	(使)通风[气]，排气	11a
versatile	[ˈvəːsətail]	*a.*	通用的，万向的，多能的	7a
versatility	[ˌvəːsəˈtiləti]	*n.*	多面性，多功能性，多方面适应性，多样性	16a
versus	[ˈvəːsəs]	*prep.*	与……比较，……对……，作为……的函数	6a
vibration signature			振动波形，振动特征	22a
virtual manufacturing			虚拟制造	30b
visual inspection			肉眼检查，目测检验	14a
vogue	[vəug]	*n.*	流行，风行	3b
volume production			批量[成批]生产，量产	27a
warp	[wɔːp]	*v.*	翘曲，变形	2a
warranty	[ˈwɔrənti]	*n.*	担保书，保证书	14a
washer	[ˈwɔʃə]	*n.*	垫圈，洗衣机，洗碗机	4a
watermill		*n.*	水车，水磨	10a
waterwheel		*n.*	水轮，水车，辘轳	10a
wear	[wɛə]	*v.*；*n.*	磨损[蚀，破，坏]，损耗(量)	5a
whilst＝while	[wailst]	*conj.*	当……的时候	29a
work envelope			工作范围	25b
work-harden			加工硬化，冷作硬化	2a
workpiece		*n.*	工件	27b
wrapping	[ˈræpiŋ]	*n.*	包(裹，装)，打包	9b
wrought	[rɔːt]	*a.*	可锻的	1b
yield	[jiːld]	*n.*；*v.*	屈服	3a
yoke	[jəuk]	*n.*；*v.*	轭，套，束缚；把……套上轭，连接	5b
zirconium	[zəːˈkəuniəm]	*n.*	锆	8a

Appendix 2 A List of Common Prefixes and Suffixes

Prefix or Suffix	Meaning or Function	Example
a-(an-)	缺,无,非	*a*typical 不典型的 *an*hydrous 无水的
ab-	脱离	*ab*normal 反常的
-able	*v.*→*a.*(表示“能……的”)	mov*able* 可移动的 work*able* 易加工的
ad-	加添	*ad*join 毗连 *ad*atom 吸附原子
aero-	空气	*aero*plane 飞机 *aero*statics 空气静力学
-age	① *v.*→*n.*	stor*age* 贮存
	② 构成抽象名词,表示“总和”	tonn*age*(船舶的)吨数 wast*age* 损耗量
-ance(-ence, ancy,-ency)	*a.*→*n.*(表示性质、程度等)	import*ance* 重要性 depend*ence* 依赖性 const*ancy* 恒定 effici*ency* 效率
-ant(-ent)	① *v.*→*a.*(……的)	resist*ant* 抵抗的 differ*ent* 不同的
	② *v.*→*n.*(……者)	result*ant* 生成物 solv*ent* 溶剂
anti-	反对,相反,抗	*anti*particle 反粒子 *anti*clockwise 逆时针方向的(地) *anti*magnetic 抗磁的
auto-	自己;自动	*auto*pilot 自动驾驶仪 *auto*control 自动控制
bi-	双,二	*bi*metallic 双金属的 *bi*atomic 二原子的
bio-	生物	*bio*electronics 生物电子学
by-	附属的,次要的	*by*-effect 副作用 *by*-product 副产品
centi-	百分之一	*centi*metre 厘米 *centi*grade 百分度的
circum-	环绕,围绕	*circum*aviate 环绕(地球)飞行 *circum*solar 围绕太阳的
co-	一起,共,和	*co*operate 合作 *co*axial 共轴的
com-	与,共,合,全	*com*munist 共产主义者 *com*pile 汇编
con-		*con*join 结合 *con*form 一致
cor-		*cor*relation 相互关系 *cor*respondence 一致
contra-	反对,相反,逆	*contra*rotation 反向旋转
counter-		*conter*clockwise 逆时针方向的(地)
de-	除去	*de*compress 减压 *de*control 解除控制
deca-(deka-)	十	*deca*metre(*deka*metre)十米
deci-	十分之一	*deci*metre 分米
di-	二	*di*oxide 二氧化物 *di*ode 二极管
dis-	① 不无	*dis*like 不喜欢 *dis*advantage 不利条件
	② 解除	*dis*integrate 分解 *dis*cover 发现
-ed	*n.*→*a.*(表示“具有……特征”)	colour*ed* 着色的 C-shap*ed* C形的
en-	① 放进,进入	*en*close 封入 *en*train 上火车
	② 使	*en*able 使能 *en*large 扩大
equi-	同等	*equi*distance 等距离
-fold	*num.*→*a.ad.*(……倍)	three-*fold* 三倍的;成三倍地
fore-	前,先,预	*fore*see 预见 *fore*cast(天气)预报
-free	*n.*→*a.*(表示“无……的”)	dust-*free* 无尘的 oil-*free* 无油的
hecto-	百	*hecto*metre 百米

续表

Prefix or Suffix	Meaning or Function	Example
hemi-	半	*hemi*sphere 半球 *hemi*cycle 半圆形
hetero-	异，杂	*hetero*-atom 杂原子
hex(a)	六	*hexa*gon 六角形
homo-	同	*homo*-ion 同离子
hydro	水	*hydro*plane 水上飞机 *hydro*statics 流体静力学
hyper-	超，过，极	*hyper*pressure 超压 *hyper*physical 超物质的
hypo	低，次	*hypo*tension 血压过低
in-(il-，im-，ir-)	不	*in*accurate 不精确的 *il*legal 不合法的 *im*practical 不实际的 *ir*regular 不规则的
infra-	在下，在下部	*infra*human 类人的 *infra*-red 红外线的
inter-	在……间；相互	*inter*planetary 行际间的 *inter*dependence 相互依赖
intro	向内；向中	*intro*nuclear 核内的
iso-	同，等，均匀	*iso*electric 等电位的
kilo-	千	*kilo*metre 千米
-less	*n.a.*(表示"没有……的")	weight*less* 失重的 stain*less* 不锈的
macro-	大，宏	*macro*molecule 大分子 *macro*climate 大气候
mal-	坏，不良	*mal*function 失灵，发生故障
meg(a)	大；兆，百万	*mega*ohm 兆欧 *mega*watt 兆瓦
-meter	度量工具	volt*meter* 伏特计 speed*meter* 测速计
micro-	微，小；百万分之一	*micro*climate 小气候 *micro*wave 微波
milli-	千分之一，毫	*milli*metre 毫米
mini-	缩，微，小(型)	*mini*-car 微型汽车 *mini*computer 微型计算机
mis-	坏，错，误	*mis*calculate 算错 *mis*apply 误用
mono-	单；一	*mono*xide 一氧化物 *mono*scope 单相管
multi-	多	*multi*phase 多相的 *multi*metre 万用表
neo-	新	*neo*realism 新现实主义
non-	非，不	*non*-metal 非金属 *non*-elastic 非弹性的
nona-	九	*nona*gon 九角形
oct(a，o)-	八	*octa*gon 八角形
out-	① 超过 ② 向外	*out*weight 重量超过 *out*grow 生长速度超过 *out*board 舷外的 *out*flow 流出
over-	超过，过分	*over*load 超载 *over*stress 过应力
pent(a)-	五	*penta*gon 五角形 *pent*ode 五极管
per-	高；过	*per*oxide 过氧化物
peri-	周围；包围	*peri*meter 周长
phono-	声音	*phono*graph 留声机 *phono*meter 测音计
phot(o)-	光	*photo*chemical 光化学的 *photo*meter 光度计
poly-	多；聚	*poly*gon 多角形 *poly*oxide 多氧化物
post-	在后；补充	*post*-war 战后的 *post*graduate 研究生
pre-	在前；预先	*pre*-college 大学入学前的 *pre*-select 预选
-proof	防……的	water-*proof* 防水的 light-*proof* 不透光的
proto-	第一，首要；原始	*proto*type 原型；范例 *proto*history 史前时期
pseud(o)-	伪，假	*pseudo*-science 假科学 *pseudo*-steel 假钢，烧结钢
quadr(i)-	四	*quadr*angle 四角形
radio-	放射，辐射；无线电	*radio*isotope 放射性同位素 *radio*-photograph 无线电照片
re-	① 再次，重复 ② 相互；返回	*re*use 重新使用 *re*train 再训练 *re*action 反作用；反应
-scope	……镜	tele*scope* 望远镜 micro*scope* 显微镜
self-	自；自身	*self*-control 自动控制 *self*-induction 自感应

续表

Prefix or Suffix	Meaning or Function	Example
semi-	半	*semi*conductor 半导体　*semi*automatic 半自动的
sept	七	*sept*angle 七角形
sex-	六	*sex*angle 六角形
sub-	① 次于;在下;低,亚	*sub*way 地下铁道　*sub*atomic 亚原子的
	② 再;分	*sub*divide 再分,细分
	③ 副,辅助	*sub*control 副控制
super-(supra-)	超	*super*conductivity 超导性　*supra*molecular 超分子的
sur-	在上,超	*sur*face 表面　*sur*pass 超过
syn-	同,共	*syn*thesis 合成
tele-	远	*tele*phone 电话　*tele*type 电传打字机
tetr(a)-	四	*tetra*gon 四角形　*tetr*ode 四极管
therm- (=thermo-)	热	*thermo*electron 热电子　*thermo*meter 温度计
-tigHT	紧密的,不透……的	water*tight* 不透水的　oil*tight* 不漏油的
trans-	横过;转移	*trans*atlantic 横渡大西洋的　*trans*mit 传输
tri-	三	*tri*angle 三角形
ultra-	超;极端	*ultra*-microscope 超显微镜　*ultra*sonic 超声(波)的
un-	① 不,未	*un*stable 不稳定的　*un*healthy 不健康的
	② 解除;废止	*un*do 松开,解开　*un*pack 启封
uni-	一,单	*uni*axial 单轴的
-ward(s)	向	for*ward(s)*向前　down*ward(s)*向下
-wise	*n.ad*.(表示"方向""位置""状态""样子")	clock*wise* 顺时针方向　length*wise* 沿着长度
with-	(相)对;逆	*with*stand 抵抗　*with*draw 撤回